MARINE BIOLOGY

MUSSELS

ECOLOGY, LIFE HABITS AND CONTROL

MARINE BIOLOGY

Additional books in this series can be found on Nova's website
under the Series tab.

Additional e-books in this series can be found on Nova's website
under the e-book tab.

MARINE BIOLOGY

MUSSELS

ECOLOGY, LIFE HABITS AND CONTROL

JAREK NOWAK
AND
MICHAL KOZLOWSKI
EDITORS

nova publishers
New York

Library of Congress Cataloging-in-Publication Data

ISBN: 978-1-62618-083-3

Published by Nova Science Publishers, Inc. † New York

CONTENTS

Preface		**vii**
Chapter 1	Native and Invasive Mussels *Slavka Stankovic and Mihajlo Jovic*	**1**
Chapter 2	Life and Interaction of Hydrobiological Organisms (Mytilus sp.) with Harmful Algal Blooms: Composition, Distribution and Metabolism of Hydrophilic and Hydrophobic Toxins in the Austral Pacific Region *Carlos García and Néstor lagos*	**37**
Chapter 3	Mussels as a Tool to Monitor Pollution *Elena Cubero-Leon and Corina M. Ciocan*	**77**
Chapter 4	Bioenergetic Model Predictions of Actual Growth and Allometric Transitions during Ontogeny of Juvenile Blue Mussels Mytilus Edulis *Poul S. Larsen, Kim Lundgreen and Hans Ulrik Riisgård*	**101**
Chapter 5	The Synthesis of Vitellogenin in the Mussel Mytilus Galloprovincialis: An Interdisciplinary Approach *Marisa Agnese, Marina Prisco, Luigi Rosati, Mariailaria Verderame, Enrico De Meo, Ermelinda Limatola, Rosanna Del Gaudio, Serena Aceto and Piero Andreuccetti*	**123**
Chapter 6	Brachidontes Solisianus (Bivalvia: Mytilidae) Re-Colonization Dynamics on A Sandstone Reef at Pernambuco - Brazil *Mariana Alves de Guimaraens, Jessica Pires Sá Carneiro, Betty Rose de Araújo Luz and Janilson Felix da Silva*	**137**
Chapter 7	Mussel Farms in the Gulf of Naples (Campania, Southern Italy): Real Opportunity or a Risk to Consumers? *Maria Grazia Esposito, Francesca Trinchella and Rosaria Scudiero*	**149**
Index		**157**

PREFACE

Mussel is the common name used for members of several families of bivalvia mollusca, from saltwater and freshwater habitats. Mussels are economically and ecologically important as they play a key role in aquatic environments and are considered to be "ecosystem engineers." In this book, the authors discuss the ecology, life habits and control of mussels. Topics include native and invasive mussels; paralytic shellfish toxins and lipophilic toxins and their interaction with the habitat of mussels in the South Pacific Ocean; mussels as a tool to monitor pollution; bioenergetic model predictions of actual growth and allometric transitions during ontogeny of juvenile blue mussels; the synthesis of vitellogenin in the mussel mytilus galloprovincialis; the Brachidontes solisiannus mussel and its re-colonization dynamics on a sandstone reef at Pernambuco, Brazil; and mussel farms in the Gulf of Naples.

Chapter 1 – Mussel is the common name used for members of several families of bivalvia mollusca, from saltwater and freshwater habitats. Mussels are economically and ecologically important aquatic mollusks. They play a key role in aquatic environments and are considered to be “ecosystem engineers”: they modify aquatic habitat, making it more suitable for themselves and other organisms. Mussels are filter feeders found all over the world. They are small, non–mobile shellfish that obtain food by filtering particulate organic matter along with indigestible components from the water column. Since they filter so much water through their bodies they are especially susceptible to water pollution.

The word “mussel” is used for many freshwater bivalves, including the freshwater pearl mussels. Freshwater mussels Unionoida inhabit permanent lakes, rivers, canals and streams throughout the world except in the Polar Regions. They require a constant source of cool and clean water. Freshwater mussels are an important food source for several different kinds of terrestrial and aquatic animals. In USA at the beginning of last century freshwater mussels were harvested and manufactured into pearl buttons, by influencing the disappearance of a large number of Unionoida species. Presently, in addition to human impacts in the past, the loss of freshwater mussels is impacted by invasive species: very small invasive freshwater mussels- zebra and quagga mussels from Europe attacked the United States lakes and rivers. Also, the golden mussel from Southeast Asia has reached rivers of South America.

The word “mussel” is also most frequently used to mean the edible bivalves of the marine mussels, family Mytilidae, most of which live on exposed shores in the intertidal zone. Marine mussels are abundant in the low and mid intertidal zone in temperate seas globally. Other species of marine mussels live in tropical intertidal areas, but not in the same huge numbers as in temperate zones. Marine mussels are eaten by humans, fishes, seabirds, and by numerous marine predators. In recent years, marine mussels have become commercially important seafood species and excellent candidate for aquaculture worldwide. Mytilidae include the well-known edible, widespread, and invasive the Mediterranean mussel, *Mytilus galloprovincialis*.

Mussels have been negatively affected by environmental changes linked to human activities, especially related to habits loss or damage, and related to organic compounds and heavy metals pollutions. The hazard of high heavy metals present in water is compounded by their ability to accumulate and biomagnifies in food chains. Much attention has recently been given to synthetic organic contaminants, which are also accumulating, in both, freshwater and marine ecosystems. Generally, mussels can serve as bioindicators of water quality and their habits. Marine mussels are recognized as good pollution bioindicators and freshwater mussels are used less as bioindicators. It appears that freshwater species may be less sensitive than their marine relatives to pollutants, in general. The zebra mussel, Dreissenidae family, is the most highly applicable bioindicator in freshwaters.

Mussels control commonly forms the basis of water monitoring, providing complete spectrum of information for proper water quality and management. They offer possibilities to obtain an ecological overview of the aquatic status. On the other hand, an invasive species is one that is not native to an ecosystem and which causes, or is likely to cause, economic or environmental harm.

Chapter 2 – For years, marine resources have been an important part of human development, given the high nutritional contribution they own. Due to requirements and needs exceeded natural inputs globally, man developed aquaculture, in order to regulate the imbalance between nutritional requirements and natural inputs, in which *Mussels sp.* are the most used species.

Under natural conditions, there are several and wide varieties of species of mussels, with multiple sea habitats: sandy, rocky and stratified bottoms in the water column in rocky strata, which allows them to assimilate a varied diet of micronutrients mostly constituted by phytoplankton. Although, depending on their habitat, mussels can filter wastes from other species such as faeces, pseudofaeces, microcellular debris or compounds from anthropogenic inputs.

From the many species of phytoplankton, about 40 of them are known by producing phycotoxins through blooms, being *Alexandrium catenella* and *Dinophysis sp.* the dinoflagellates'species associated with the production of toxins in the South Pacific Ocean. These dinoflagellates are filtered and accumulated by hydrobiological organisms, such as mussels, that accumulate toxins in the digestive glands, where are subjected to numerous reactions of biotransformation mediated by enzymatic, nonenzymatic pathways or by bacteria present in bivalves, which extends the toxic spectrum of the assimilated toxins, an event that is magnified by the interaction of vector organisms that expand the toxic forms through the trophic chain.

Paralytic shellfish toxins (STX isomers) and lipophilic toxins (DSP-toxins, AZAs, PTXs and YTXs), constitute the group of toxins that affect mussels from natural origin and aquaculture the most. These kind of toxins are accumulated in the tissues of mussels where are retained according to their chemical affinity and modified into more toxic forms (C2→GTX3) and/or thermodynamically more stable forms (GTX3→GTX2; 7-epi-PTX-2-

SA→PTX-2-SA), allowing a harmlessness to accumulated toxins by mussels in some cases (DTX1→DTX3; AZA3→AZA5).

Since blooms associated with PSP-toxins and lipophilic toxins may occur simultaneously, different species of mussels and those species related to their habitat can accumulate and/or transform the wide spectrum of toxins, generating a toxin profile and variable toxicity in their tissues, which last according to the exposure time to the bloom and the purification pathway of toxins from mussels.

In this paper, the main events associated with paralytic shellfish toxins and lipophilic toxins as well as their interaction in the habitat of *Mussels sp.* in the South Pacific Ocean are presented.

Chapter 3 – In order to evaluate the impact and potential risk that toxicants have in the environment, biomonitoring programs are used. In general, they can fall into two categories, exposure and effects assessment. The traditional exposure assessment process involves the analytical measure of a target compound in the tissue of a sampled organism or in the water. Although exposure does not necessarily mean effects in the organisms, it can assist in confirming that the effects seen are due to anthropogenic impacts and not to natural alterations. Effects monitoring, on the other hand, looks at various levels of the biological organization to evaluate the status of the biological community. Mussels have been used as sentinel organisms to monitor effects and exposure to a variety of chemical contaminants in the aquatic environment such as metals, organochlorines, pesticides and PAHs and are being used in biomonitoring studies from international programmes like "Mussel Watch". Their sessile nature, wide geographical distribution, large populations, high filtering rates, and poor ability to metabolise xenobiotics make them ideal organisms for bioaccumulating contaminants from the aquatic environment and make them suitable as models in the field of environmental toxicology. This chapter focuses on the current status and perspectives on their usage as biological indicators to monitor pollution.

Chapter 4 – The growth rates of blue mussels *Mytilus edulis* on ropes in the Great Belt (Denmark) have been studied during the growth season of one year, from settling to about 30 mm shell length mussels, covering > 4 decades of body mass. Measured shell length (L, mm) and dry weight of soft parts (W, µg) for $L > 10$ mm followed a power-law ($W = 2.15L^{3.40}$) which supplemented an existing power-law for $L < 10$ mm ($W = 24.7\,L^{2.42}$) to establish that somatic growth changes character at $L \approx 10$ mm and $W \approx 10$ mg. Results of specific growth rates based on dry weight of soft parts ($\mu = \mathrm{dln}W/\mathrm{dt}$) compared well with predictions based on a previously developed bioenergetic growth model (BEG) for $W > 10$ mg ($\mu = \mathrm{a}W^{b}$, $a = 0.871\times C - 0.986$; $b = -0.34$, with µ in % d^{-1} and W in g) which explicitly takes into account the prevailing chl *a* concentration C (µg L^{-1}). Results for $W < 10$ mg also correlated well by the power-law ($\mu = \mathrm{a}W^{b}$), now with exponent b = −0.13 close to the suggested value (b ≈ −0.1) from experimentally established correlations for filtration and respiration rate of post-metamorphic mussels. Using the stated $W(L)$-relation for $L > 10$ mm the growth model has been expressed in terms of shell length specific growth rate ($\mu_L \equiv \mathrm{dln}L/\mathrm{dt} = \alpha L^{\beta}$) by which data on shell length was well correlated, including the influence of chl *a* concentration. Supplementary growth data from mussels in suspended net-bags at the same site illustrated differences ascribed to lack of competition for space and food, and literature data on shell length from cage-growth of mussels in the brackish Baltic Sea support the present correlations. It is argued that the allometric transitions that take place around $W \approx 10$ mg and

$L \approx 10$ mm during the ontogeny of *M. edulis* is most likely universal and not restricted to first year growth of juvenile (young) mussels during the productive season.

Chapter 5 – Vitellogenins (VTGs) represent the most important precursors of vitellins (Vns), the egg yolk proteins utilized by oviparous and ovoviviparous animals to guarantee the embryo development. In mature females, VTGs are synthesized in different districts under the control of sexual hormones, as 17β-estradiol, and transferred to the growing oocytes via the bloodstream. In the oocytes, VTGs are processed by cathepsins into Vns that are then stored into the yolk platelets. Differently, in immature females and in the males VTGs are not synthesized, but their synthesis may be activated by xeno-estrogens, an heterogeneous group of environmental contaminants that modify the endocrine reproductive system with consequent production of molecules not present physiologically, as VTG. For this reason, VTG production, in immature females and males, is an useful biomarker to study the health of aquatic and terrestrial ecosystems. Till now, the VTG was utilized as biomarker mostly in fish, and more recently in mussels too. Such organisms indeed have acquired a prominent role as sentinel species in aquatic pollution-monitoring programs. Differently from fish, the knowledge of VTG synthesis and its endocrine control in the mussels is scant. At this regard the authors carried out an interdisciplinary investigation to identify the site/s of VTG synthesis in the mussel *Mytilus galloprovincialis*, that may represent a useful sentinel species in the Mediterranean sea. In this organism the authors have investigated two districts, the ovary and the digestive gland, that are the most probably sites of VTG synthesis in the mussels. Results of RT-PCR, in situ hybridization, electron-microscopy and immunohistochemistry experiments, taken together, provide evidence that VTG synthesis occurs only in the ovary of females, and particularly both in the oocyte (autosynthesis) and follicle and adipogranular cells (eterosynthesis). The possibility that in the mussel *Mytilus galloprovincialis* VTG may be used as biomarker in ecotoxicology is also discussed.

Chapter 6 – Three sub-environments were identified on the reef flat at Piedade Beach. Each sub-environment was characterized by different organisms on the intertidal zone. The mesointertidal contained the algal sub-environment, while the mussel *Brachidontes solisianus* (d´Orbigny, 1846) dominated the bivalve sub-environment, and patches of Cirripedia dominated the higher intertidal. Mussels were collected by scraping, three quadrats in each sub-environment each month. The mussels were classified in three size classes according to shell length. Each month´s sample was counted and measured to evaluate re-colonization patterns and correlations with environmental factors, such as air temperature, rainfall and average algal biomass, within distinct sub environments between March and December of 2007. The field data show a positive relationship between size classes in the *Brachidontes solisianus* colonization experiments in the Cirripedia, mussel and algal sub-environments. These results suggest that mussel recruitment tends to occur closer to older individuals, although the field observations indicated that the patch closure was the principal process of recovery from disturbance. A positive correlation was found between rainfall and mussel recruitment in the Cirripedia sub-environment, and a positive relationship between bivalve recruitment and macroalgal biomass was found in the *Brachidontes* and algal sub environments. A covariance analysis showed significant differences in bivalve density among the three sub-environments. *Brachidontes* recruitment was highest in October for the algal sub-environment (5483 individuals/m^2), whereas colonization was highest in June 2007 in the invertebrates sub- environments.

Chapter 7 – In recent years, in Italy it has been observed a process of post-industrialization. Many industries closed, and at the same time is occurring a return to more environmentally friendly practices, such as organic farming and responsible fisheries. For the latter the decline of wild aquatic organisms has led to the expansion of aquaculture. Among the molluscs, the edible mussel *Mytilus galloprovincialis* is the most sold in the Southern Italy, and the animals come almost exclusively from aquaculture. The Gulf of Naples is a 15 km coastal area located in the Campania region (Southern Italy), subject to pollution from highly populated urban centers, port activities, and large industrial complexes, including oil refineries. Nevertheless, nowadays many mussel farms have been arising in this area following the closure of many important industries (the steel plant ILVA, for example). This chapter summarizes the results obtained by monitoring the quality of the mussels *M. galloprovincialis* farmed in different areas of the Gulf of Naples, in terms of heavy metal contamination. The concentration of toxic heavy metals (Pb, Cd) was determined on the tissues of *M. galloprovincialis* specimens collected at different aquaculture farms. On the same animals was also determined the amount of metallothionein, the most important biomarker of metal pollution. The collected data were compared with those obtained from animals belonging to farms located in pristine areas and from animals relaying in metal free seawater, in order to establish public health risks associated with consuming mussels harvested in the Gulf of Naples.

In: Mussels: Ecology, Life Habits and Control
Editors: Jarek Nowak and Michal Kozlowski
ISBN: 978-1-62618-083-3

Chapter 1

NATIVE AND INVASIVE MUSSELS

Slavka Stankovic*[*,1] and Mihajlo Jovic*[2]
[1]Faculty of Technology and Metallurgy, University of Belgrade, Belgrade, Serbia
[2]Vinca Institute of Nuclear Sciences, University of Belgrade, Belgrade, Serbia

ABSTRACT

Mussel is the common name used for members of several families of bivalvia mollusca, from saltwater and freshwater habitats. Mussels are economically and ecologically important aquatic mollusks. They play a key role in aquatic environments and are considered to be "ecosystem engineers": they modify aquatic habitat, making it more suitable for themselves and other organisms. Mussels are filter feeders found all over the world. They are small, non–mobile shellfish that obtain food by filtering particulate organic matter along with indigestible components from the water column. Since they filter so much water through their bodies they are especially susceptible to water pollution.

The word "mussel" is used for many freshwater bivalves, including the freshwater pearl mussels. Freshwater mussels Unionoida inhabit permanent lakes, rivers, canals and streams throughout the world except in the Polar Regions. They require a constant source of cool and clean water. Freshwater mussels are an important food source for several different kinds of terrestrial and aquatic animals. In USA at the beginning of last century freshwater mussels were harvested and manufactured into pearl buttons, by influencing the disappearance of a large number of Unionoida species. Presently, in addition to human impacts in the past, the loss of freshwater mussels is impacted by invasive species: very small invasive freshwater mussels- zebra and quagga mussels from Europe attacked the United States lakes and rivers. Also, the golden mussel from Southeast Asia has reached rivers of South America.

The word "mussel" is also most frequently used to mean the edible bivalves of the marine mussels, family Mytilidae, most of which live on exposed shores in the intertidal zone. Marine mussels are abundant in the low and mid intertidal zone in temperate seas globally. Other species of marine mussels live in tropical intertidal areas, but not in the

[*] Tel: + 381 11 3303 778; Fax: + 381 11 3370 387; Email: slavka@tmf.bg.ac.rs.

same huge numbers as in temperate zones. Marine mussels are eaten by humans, fishes, seabirds, and by numerous marine predators. In recent years, marine mussels have become commercially important seafood species and excellent candidate for aquaculture worldwide. Mytilidae include the well-known edible, widespread, and invasive the Mediterranean mussel, *Mytilus galloprovincialis*.

Mussels have been negatively affected by environmental changes linked to human activities, especially related to habits loss or damage, and related to organic compounds and heavy metals pollutions. The hazard of high heavy metals present in water is compounded by their ability to accumulate and biomagnifies in food chains. Much attention has recently been given to synthetic organic contaminants, which are also accumulating, in both, freshwater and marine ecosystems. Generally, mussels can serve as bioindicators of water quality and their habits. Marine mussels are recognized as good pollution bioindicators and freshwater mussels are used less as bioindicators. It appears that freshwater species may be less sensitive than their marine relatives to pollutants, in general. The zebra mussel, Dreissenidae family, is the most highly applicable bioindicator in freshwaters.

Mussels control commonly forms the basis of water monitoring, providing complete spectrum of information for proper water quality and management. They offer possibilities to obtain an ecological overview of the aquatic status. On the other hand, an invasive species is one that is not native to an ecosystem and which causes, or is likely to cause, economic or environmental harm.

Keywords: Marine, Freshwater, Invasive, Mussels, Pathogenic bacteria, Toxic metals, Organic compounds

List of Abbreviations

cfu	colony forming units
CLRTAP	Convention on Long-Range Transboundary Air Pollution
DDT	Dichlorodiphenyltrichloroethane
EC	European Commission
MarLIN	Marine Life Information Network
N	Nitrogen
NOAA	National Oceanic and Atmospheric Administration
NRCS	Natural Recourses Conservation Service
NSandT	National Status and Trends
P	Phosphorus
PAH	Polycyclic Aromatic Hydrocarbons
PBB	Polybrominated Biphenyls
PBDE	Polybrominated Diphenyl Ethers
PCB	Polychlorinated Biphenyl
PCDD/Fs	Polychlorinated Dibenzodioxins / Dibenzofurans
POP	Persistent Organic Pollutants
RNO	Reseau National d'Observation de la qualité du milieu marin
TBT	Tributyltin
UK	United Kingdom
UNDP	United Nations Development Program

UNECE	United Nations Economic Commission for Europe
UN/IOC	United Nations Intergovernmental Oceanographic Commission
USA	United States of America
USFWS	U.S. Fish and Wildlife Service
WHO TEQ	World Health Organization Toxic Equivalent

1. Introduction

Mussels are found in all types of water including brackish, marine and fresh water (Mozley 1954). They live from the Arctic to the Antarctic, in oceans and in brooks, rivers, ponds and lakes on every continent. They flourish in tropical jungles and on sandy beaches as well as in shallow lagoons and water as deep as 6000 m (Rehder 1981). Their specialized anatomy, physiology and morphology allow them to thrive in diverse water environments. Each species is uniquely adapted to their endemic conditions.

The word "mussel" is most frequently used to mean the edible bivalves of the marine family Mytilidae, most of which live on exposed shores in the intertidal zone (Stankovic et al. 2011, Stankovic and Jovic 2012). The word "mussel" is also used for many freshwater bivalves, including the freshwater pearl mussels. Freshwater mussel species inhabit bottoms of lakes, ponds, rivers, creeks, canals, grouped in a different subclass, despite some very superficial similarities in appearance (Helfrich et al. 2009).

Freshwater mussels are bivalves belonging to the Order *Unionoida*. Unionids are among the longest lived animals, with most species living for decades and some to over a century. The combination of complex life cycle, exposure to the vertical range of aquatic and benthic environments, species diversity and range, ecological importance, and widespread imperilment of native mussels make this taxon uniquely attractive for the study of water and sediment pollution effects (Barnhart and Neves 2005). Freshwater mussels are often the dominant component of the invertebrate benthos in streams and are ecologically significant as suspension feeders.

Environmental pollution has increased substantially in the last decades due to a great number of industrial, agricultural, commercial and domestic waste, effluents and emissions as well as hazardous substances. The majority of environmental pollutants are threatening initially human and environmental health, but also the integrity and functions of ecosystems. Aquatic pollution is the predominant form of pollution, since the majority of chemical pollutants are entering wetlands, rivers, lakes and seas (Stankovic et al. 2011, Stankovic and Stankovic 2012).

Different type of water pollution becomes a vital problem in many countries worldwide. This negative phenomenon is difficult to control or eliminate, and therefore, the issue requires prompt solutions. The function of water as a receiver of polluted runoff water from land implies that the use of this water for various purposes (e.g. drinking water, agriculture, aquaculture, recreations) might need operations for its quality improvement (Brzozowska et al. 2012). Pollutions are present in waters, in sediments, and in the tissues of its living organisms, including mussels. Toxins like heavy metals and organic contaminants can pose a threat to the health of aquatic life and humans (Stankovic and Stankovic 2012).

Pollutants, such as heavy metals and organics, are released into the environment—the surface water and air, hence into sediments and seawaters: once in the environment, they do not disappear, but accumulate in freshwaters, seawaters, sediments and biomass. Pollutants in mussels depend not only on their concentrations in water and sediment, but also depend on environmental conditions, biological factors of the organisms themselves. It is therefore necessary to monitor the impact of all these factors in this complex system (Stankovic and Stankovic 2012). Additionally, early life stages of freshwater mussels are known to be especially sensitive to certain contaminants, such as ammonia, metals, and some organics and aromatic hydrocarbons relative to other aquatic species (Bringolf et al. 2007). Also, mussels which inhabiting polluted waters are able to concentrate micro–organisms contaminated by human pathogens and therefore are used as indicators of their presence in the aquatic environment and to predict potential health concerns transmitted by water or mussel consumption (Graczyk et al. 2008).

In last two decades several non–indigenous bivalve species have been colonizing aquatic ecosystems worldwide, in some cases with great ecological and economic impacts, (Darrigran 2002, Sousa et al. 2009, Stankovic et al. 2011), like the golden mussel (*Limnoperna fortunei*) in South America, South Europe and Japan and the Mediterranean mussels (*Mytilus galloprovincialis*) in Asia, North America and South Africa. The continued invasion of freshwater mussels like zebra mussels (*Dreissena polymorpha*) and quagga mussels (*Dreissena rostriformis bugensis*) in North America and Europe has threatened the survival of native Unionid mussels.

Many invasive mussels occur at high densities and over a large spatial extent. Mussels naturally filter an enormous amount of water daily, and steadily remove pollutants, micro particles, and organic compounds from the water (Baker and Levinton 2003). This ecological service is free of charge to humans. But in addition, changes in ecosystem structure and functioning due to engineering by invasive mussels often have very large ecological and economic impacts. The worldwide spread of non–indigenous mussels and the varied ways in which they physically modify habitats, their engineering effects should receive more serious consideration (Sousa et al. 2009).

At the first moment, the fact that many invasive mussel species have positive effects on invertebrate density, biomass and species richness does not necessarily mean that they exclusively have beneficial effects on other species. Increases in total invertebrate density and richness after the invasive mussels are usually associated with decreases, and sometimes, extirpation of particular species (Darrigran 2002). Therefore, it is to be expected that their engineering activities can significantly alter ecosystem structure and functioning, also causing economic impacts (Sousa et al. 2009). Understanding their ecosystem roles is crucial for their control.

2. Mussels

Classification

Mussels form a large family and live in both, saltwater and freshwater habitats. Mussels are shell–covered invertebrate animals and share the taxonomi Mollusca with oysters, snails, octopi, and squid, among others. Most mussels are filter–feeders, but can be parasitic during their larval stages. While naming scientific mussels (taxonomy) is an exact science, they are known by many common names within the class Bivalvia as: clam, oyster, bivalve, mollusk, unionid, scallop, shellfish, and seashell, and of course, mussel.

There are eight classes of mollusks, which are separated primarily based on differences in the feet and shells. Clams, mussels and oysters belong to the class known as Bivalvia, which indicates that there are two laterally posed, symmetrical valves (shells). These valves are opened by an elastic ligament near the hinge and closed by adductor muscles, particularly when threatened.

Mussels have several subclasses: Pteriomorpha (marine mussels), Palaeoheterodonta (freshwater mussels) and Heterodonta (zebra mussels). Marine mussels have in common a shell whose outline is elongated and asymmetrical compared with clams, which are often more or less rounded or oval. Those that live in rivers and fast moving streams generally have heavier shells to help prevent being washed away by the current, while those in pools only need shells thick enough to deter predators. Zebra mussels are in the Dreissenidae family, the false mussel and zebra mussel family. The similarities of morphology, size, and behavior between freshwater zebra mussels and marine bivalves are strong (Ackerman et al. 1994). Freshwater Zebra mussels and their relatives in the family Dreissenidae are not related to marine mussels, even though they resemble many *Mytilus* species in shape, and live attached to rocks and other hard surfaces in a similar manner, using a byssus. It is invasive species, occurs at a high abundance, widely distributed, and continues to spread to different catchments. They are classified with the Heterodonta, the taxonomic group which includes most of the bivalves commonly referred to as “clams”.

Anatomy

The visceral mass of mussels contains muscles for digestion, reproduction, circulation and respiration. Two external tissue flaps, known as the mantle, enclose the mantle cavity and secrete the substance that produces the shell.

Specifically, bivalves like clams, mussels and oysters have hatchet–shaped feet, a pair of gills each and mantle lobes next to each shell. A bivalve's back end has two tubes or siphons, one for taking water in and one for expulsing it. The “intake” siphon receives oxygen and food, while the “outtake” siphon eliminates waste. Clams have well–developed feet for burrowing, but mussels and oysters have relatively weak feet. Unlike other mollusks, only bivalves and a few others in the phylum lack a radula, a rasping organ used for specialized feeding techniques (Helfrich et al. 2009).

Pearl Producers

Bivalves like clams, oysters and mussels make pearls. Pearl forms when an irritant embeds itself into a bivalve's mantle. In a defensive measure, the animal covers the irritant in the same substance it uses to create the lining of its shell, which is smooth and luminescent. Over time, the irritant is covered in multiple layers and forms the center of a sphere; this thickly coated irritant is known as a pearl. Many species are utilized as sources of mother–of–pearl (Helfrich et al. 2009).

Interaction with Humans

Clams, mussels and oysters have long been used by humans in agriculture, for food and to demonstrate or establish wealth. When mollusk shells are crushed, they attract microorganisms. Those microorganisms, in turn, kill nematodes that plague agricultural fields. Aesthetically pleasing shells are prized for their rarity and beauty, and pearls have long been a symbol of wealth and stature (particularly the rare, naturally formed, nearly perfect spherical examples). Some early cultures also used shells like those from clams, mussels and oysters as a form of currency.

Humans have used mussels as food for thousands of years and continue to do so. About 17 species are edible, of which the most commonly eaten are *Mytilus galloprovincialis, Mytilus edulis, Mytilus trossellus* and *Perna canaliculus*. Marine mussels have a high nutritive value as a result of their complex compositions, rich in biological active components: essential amino acids, hydrosoluble and liposoluble vitamins, fatty essential acids, microelements (Stankovic and Jovic 2012).

Freshwater mussels nowadays are generally considered to be unpalatable, though the native peoples in North America ate them extensively. During the Second World War in the United States, mussels were commonly served in diners. This was due to the unavailability of red meat related to wartime rationing.

Historically, freshwater mussels were very common, and they were most often harvested for button making and their soft tissues for livestock feed in North America. The button industry declined with the advent of plastics in the 1950s. By the 1950s, however, the Japanese had found a new market for freshwater mussels as a source material for cultured pearls.

Nowadays thousands of tons of mussel shells are processed and exported to Japan each year to supply the cultured pearl industry. Additional commercial and medical uses for freshwater mussels are under consideration (NRCS 2007).

Pollution Monitors

Mussels, like other filter feeders, tend to accumulate toxins and pollutants. Because of this, a sampling of them is often related to certain regions to gauge water and sediment pollution levels, (Stankovic et al. 2011, Stankovic and Jovic 2012, Stankovic and Stankovic 2012). Although mussels are valued as food, mussel poisoning due to toxic planktonic organisms can be a danger along some coastlines. For instance, mussels should be avoided along the west coast of the United States during the warmer months. This poisoning is usually due to a bloom of dinoflagellates (red tides), which contain toxins. The dinoflagellates and their toxin are harmless to mussels, even when concentrated by the mussel's filter feeding, but if the mussels are consumed by humans, the concentrated toxins cause serious illness, such as paralytic shellfish poisoning. Usually the U.S. government monitors the levels of toxins throughout the year at fishing sites (Nalepa 2010).

Nonnative Aquatic Mussels

Although species distribution changes naturally over time, human activities greatly increase the rate and the spatial scale of these changes by accidentally or deliberately moving organisms across the world. The impact of the nonnative mussel is not only restricted to economic aspects: pipe obstruction; reduction in flow velocity in pipes due to friction loss, accumulation in empty valves and the pollution of water ways from massive mortality; filter occlusion and an increase in the corrosion of surfaces due to mussel infestation. Beside these new economic unexpected expenses, such as system shutdowns, pipe and filter replacement, the need for chemical or mechanical cleaning, also environmental problems are produced. The rapid change in benthic communities should be noted: the large biomass associated with high densities of invasive mussels may impact aquatic food chain and habits (Darrigran 2002).

According to Darrigran (2002) an invasive species should have several characteristics to be successful in the new environment: a short life span (from 2 to 3 years); rapid growth; rapid sexual maturity; high fecundity; an ability to colonize a wide range of habitat types; possessing a wide range of physiological tolerances; gregarious behavior; some form of association with human activities (inhabitants of estuaries, rivers, lakes or canals, reservoirs, harbors, etc.); suspension feeding; the ability to repopulate previously colonized habitats, following population crashes caused by extreme physical conditions. Mediterranean mussels (*Mytilus galloprovincialis),* and freshwater mussels, the zebra mussel (*Dreissena polymorpha*), the quagga mussel (*Dreissena rostriformis bugensis*) and the golden mussel (*Limnoperna fortunei*), as invasive mussels, present many of the above characteristics (Ludyanskiy et al 1993, Darrigran 2002).

Natively the zebra mussel inhabits parts of western Russia, near the Caspian Sea and Ural River. From its native origin the species has spread to the Eastern Europe. It took approximately 200 years for the zebra mussel to spread from its origins to the rest of western Russia and all of Europe. The first zebra mussel sighting in the United States was in 1988 in Lake St. Clair. To occupy North America, the evidence suggests this invasion is happening much more quickly than the mussel's spread across Europe (Ludyanski et al. 1993).

According to Sagoff (1999) industries have to take expensive steps to keep these creatures from colonizing intake pipes used for water works and power plants. On the other side, the zebra mussel, a filter feeder, is credited with clearing the water column of nutrients excess and associated algae resulting from municipal waste discharge and agricultural runoff. Lake Erie, which had once been given up as dead by eutrophication, is now clear of the organic matter that had been choking it, wholly because of the mussel. The zebra mussel gathers excess of nutrients into particles depositing it at the bottom of lakes and rivers to form excellent habitat for insect larvae, leeches, snails, and other invertebrates that larger fish, such as yellow perch, feed upon. The work of this mussel performs in clearing the water column and enhancing benthic invertebrate communities in Michigan's Lake St. Claire (Sagoff 1999). Sagoff (1999) concluded that the benefits of zebra mussels are ignored however, because it is an "alien" species.

Like the zebra mussel, the quagga mussel (*D. bugensis*) is believed to have been transported from Europe waters and was established in the Great Lakes by the 1980s. Quagga mussels are very similar to zebra mussels; however, they can survive in deeper water and reproduce at lower temperatures. Zebra and quagga mussels not only affect native mussels, but also their ability to rapidly colonize hard surfaces causes serious economic problems. These organisms can clog water intake structures, such as pipes and screens, thereby reducing pumping capabilities for power and water treatment plants, costing industries, companies, and communities. Recreation based industries and activities have also been impacted, and beaches have all been heavily colonized (NRCS 2007).

The quagga mussel (*D. bugensis*), originally are from the Ukraine, infested the Great Lakes region competing with their earlier arrived relative, the zebra mussel (*D. polymorpha*). These mussels have a high potential to adapt in new environmental conditions (Mills et al. 1996). They are currently of major concern in the United States as an invasive species (USFWS 2007).

Limnoperna fortunei, the golden mussel, is a freshwater bivalve native to Southeast Asia that has become a pest species of global concern over the last few decades. Between 1965 and 1990 it was introduced into Hong Kong, Taiwan, Japan and South America, most probably via ship ballast water (Boltovskoy et al. 2009). The golden mussel (*L. fortunei*) was introduced into Argentina from Asia in 1991 and within a decade spread in rivers of four other South American countries and Mexico Golf as a dominant component of the benthic fauna and a major nuisance for industry and power plants (Darrigran 2002). The golden mussel exhibits a wider tolerance of ecological parameters than the zebra mussel, and therefore, the golden mussel is expected to invade a broader range of habitats (Magara et al. 2001).

2.1. Life Habits

The first difference between fresh and marine mussels is the salinity tolerance. Some mussels can survive in the intermittent estuarine waters, while others cannot. All mussels have internal osmoregulation that maintains their water content, pH, and salinity levels. If an organism or cell cannot regulate its internal conditions, such as the chemical composition of its body fluids, so as to maintain health and functioning, regardless of outside conditions, then the mussel cannot survive (McMahon and Bogan 2001).

2.1.1. Native Freshwater Mussels

Distribution and Habitat

Freshwater mussels belong to the phylum Molluska, the second most diverse group of animals in the world in terms of number of described species. The phylum consists of approximately 100,000 freshwater, marine, and terrestrial species (NRCS 2007). Freshwater bivalves are one kind of freshwater mollusks. They are bivalves which live in freshwater, as opposed to saltwater, the main habitat type for bivalves. The majority of species of bivalve mollusks live in the sea, but in addition, a number of different families live in freshwater, and in some cases also in brackish water.

Freshwater mussels (Order Unionoida) are among the most widespread, diverse, and historically abundant animals in freshwaters. The Unionoida is, in turn, divided among six families: Unionidae, Margaritiferidae, Hyriidae, Etheriidae, Mycetopodidae and Iridinidae. These families have been well–known historically, although often by other names.

All freshwater mussels belong to the mollusk order *Unionoida*, a solely freshwater group. These animals are usually from 15 mm up to 15 cm in length. They live in many types of habitat, ranging from small ditches and ponds, to canals, rivers and lakes. Species from those groups vary greatly in size: some of the pea clams (*Pisidium* species) have an adult size of only 3 mm, and in contrast, one of the largest species of freshwater mollusk order *Unionoida* is the swan mussel. It can grow to a length of 20 cm, and usually lives in lakes or slow rivers. The *Unionoida* are the pearly freshwater mussels. All reproduce by means of a larval stage that is parasitic on fish. Freshwater pearl mussels are economically important as a source of freshwater pearls and mother–of–pearl, (NRCS 2007).

The freshwater pearl mussel is distributed from the Arctic and temperate regions of western Russia through Europe to the northeastern seaboard of North America. Recent studies have revealed that there have been dramatic declines throughout its Holarctic range: for example, Bauer (1988) estimates a 95–100 % decline in known populations in central and southern Europe. Substantial populations with active recruitment are now found on fewer than 50 rivers in Canada, northwest Russia and northeast Scandinavia, with a handful of sites in Bavaria, the Czech Republic and Austria. The early causes of decline were undoubtedly pearl fishing and industrial pollution (Skinner et al. 2003).

The freshwater pearl mussel is found throughout northwest Europe, but in declining numbers. The UK is home to 6 native species of freshwater mussel of which two, the depressed river mussel (*Pseudanodonta complanata*) and the pearl mussel (*Margaritifera margaritifera*), have experienced significant declines during the 20th century. Both are now included on a list of the most threatened UK species drawn up as part of the Government's Biodiversity Action Plan (Skinner et al. 2003).

Freshwater mussels are found living on the bottoms of lakes and streams throughout the world. About 1,000 species have been identified worldwide, and 300 (nearly 30 %) of these occur in North America, 96 in Africa, 60 in China and 12 native species in Europe (NRCS 2007). Historically, freshwater mussels were very important to Native American tribes, since they often used them in tribal religious ceremonies and even for food (Garvin 2005). The abundance and variety of native freshwater mussels in North America have declined sharply in the last century, (Helfrich et al. 2009).

Although freshwater mussels are distributed throughout the world, they reach their greatest diversity in North America, east of the Mississippi River. In the early 1900s millions of mussels were harvested from rivers in North America to make pearl buttons out of the shells. With the advent of cheaper and more durable plastic buttons, the mussel mother–of–pearl button industry died. However, since the late 1950s a new demand for mussel shell originated – to supply the Japanese cultured pearl industry (Helfrich et al. 2009).

About 300 freshwater native species are found in North America, with <10 species per state in the West to over 100 species per state in the Southeast of USA. Unfortunately, much of this diversity is threatened by a variety of human impacts and as a consequence approximately 67 % of the nearly 300 native freshwater mussel species are considered vulnerable to extinction and already at least 35 species are extinct, and more extinction is impending (NRCS 2007). Nonetheless, native mussels are a dominant component of the benthic community in US rivers, with peak abundance reaching 100 individuals and 10 kg per m^2 (Barnhart and Neves 2005).

In summary, native freshwater mussels are a highly diverse, historically abundant and widespread group. They are ecologically important and, as a group, are at a critical stage for conservation because of their large number of imperiled species. Freshwater mussels inhabit permanent lakes, rivers, canals and streams throughout the world, except in the Polar Regions. They require a constant source of cool, clean water and non muddy bottom. Mussels typically inhabit unpolluted, standing or flowing waters that are rich in oxygen, calcium, and suspended food particles (Helfrich et al. 2009). They prefer water with a substantial mineral content, using calcium carbonate to build shells (NRCS 2007). These mussels are used, and still using, as host animals for the cultivation of purls.

Feeding

Freshwater mussels are often found in aggregations called mussel beds, which can be a mile or more long and contain thousands of mussels. The water over a pearl mussel bed must be clear and well oxygenated. Maximum depth is about 2 m (NRCS 2007). Mussels feed primarily through filtration of the water column or the sediment pore water. Food sources include detritus, bacteria, phytoplankton and zooplankton. Freshwater mussels may deposit and/or suspension feed; data suggesting both feeding methods exist. Filtration rates may approach 6 L/hour.

They are filter– and deposit–feeding, long–lived (40–100 years) organisms that live burrowed in sediments of streams and rivers. Adult mussels are suspension feeders that live partially or completely buried in sediments. Mussels filter prodigious volumes of water, capturing particles down to 5 microns in size and ingesting algae, zooplankton, bacteria and detritus. They are therefore intimately exposed to both dissolved and suspended contaminants. Young juvenile mussels are very small (<400 microns) and occupy interstitial spaces in sediments where they ingest both deposited and suspended particles. At this stage, juveniles are likely exposed primarily to contaminants in pore water and associated with sediment particles. As they grow, most species occupy the water–sediment interface and draw water mainly from the water column. However, adults may also filter pore water when they burrow below the surface (Barnhart and Neves 2005).

Adults bury themselves in the bottom sediment with a fleshy muscular foot and live by filtering algae and other food items from the water column. The freshwater pearl mussel life cycle involves an adult stage, living as a filter feeder, a juvenile stage, living interstitially in sediment, and a larval (glochidial) stage living attached to the gills of trout or salmon. All life stages therefore need consideration, as does the viability of the host species of fish (Barnhart and Neves 2005).

The freshwater pearl mussel is one of the longest–lived invertebrates known, and individuals can survive for over 100 years, reaching 12–15 cm in length. The life span and maximum size reached are highly variable between populations, depending on environmental conditions, particularly hydrochemistry and water temperature (Skinner et al. 2003). The mussels live buried or partly buried in coarse sand and fine gravel in clean, oligotrophic, fast–flowing and unpolluted rivers and streams. They inhale water through their exposed siphons to filter out minute organic particles on which they feed. Where the species was formerly abundant, it is possible that this filtration acted to clarify river water to the benefit of other species, including juvenile Atlantic salmon (*Salmo salar*) and brown or sea trout (*Salmo trutta*) (Zuiganov et al. 1994).

Reproduction

Most mussels reproduce annually. Reproduction in freshwater mussels occurs when the male–released spermatozeugmata are captured by the females, and the eggs are fertilized. Developing eggs are brooded in the gills of the female, and eventually develop into glochidia, the larval stage of the mussel. The glochidia, which have "teeth" that facilitate attachment to the host fish, are then released by the female mussel and attach to the gills of a host fish species. Glochidia survival time after being released from the female mussel ranges from less than two days to nearly 10 days. Once on the fish host, the glochidia are encysted by host tissue. Host fish suitability depends on the mussel species. Some mussel species have only one suitable host fish while others may have many. Atlantic salmon and brown trout are the main hosts for pearl mussel glochidia. Therefore, measures to conserve the mussel must also include the host fish too (Skinner 2003).

The attached glochidia develop into juveniles, which eventually exist and fall into the substrate and develop into adult mussels, which are often long–lived. Generally, growth rates are most dramatic in the first years of existence and then decline over time. Every life stage of the mussel may be sensitive to pollutants, particularly the long–lived adult stage and the vulnerable larval stage (Barnhart and Neves 2005). In some rivers, longevity may also be reduced where surrounding limestone strata increases calcium input (Skinner 2003).

The freshwater pearl mussel is matures at 10–15 years, when the length generally exceeds 65 mm. In early summer (June to July), the males shed sperm into the water, and it is inhaled by the females. The fertilized eggs develop in a pouch on the gills for several weeks, and are released from July to September as tiny larvae, known as glochidia. Each female ejects between 1 and 4 million glochidia, usually over one to two days. The proportion of adults producing glochidia is relatively high, varying from 30–60 %, even in sparse populations. Almost all the glochidia are swept away and die, but a few are inhaled by juvenile Atlantic salmon and brown or sea trout and most attachments occur within a few hours. Glochidia attached to the gills of juvenile fish encyst, live and grow in the hyper–oxygenated environment until the following spring. They drop off in May and early June, and must land in clean, sandy or gravelly substrates to settle and start to grow. Growth to sexual maturity requires approximately 2–6 years (Barnhart and Neves 2005). The huge losses involved in this unusual life cycle make the freshwater pearl mussel particularly vulnerable to adverse conditions (Skinner et al. 2003). However, mussel distributions are dependent on the distribution and abundance of their fish hosts (Zanatta and Murphy 2006).

Mussels have a vulnerable life history because: (1) the long lived adults require stable benthic habitats in rivers; (2) they are filter feeders and therefore are exposed to water–borne pollutants; (3) they have a high–risk parasitic larval stage that are dependent on native fish hosts and (4) the tiny juvenile mussels are susceptible to drift, siltation, micro predators and sediment toxicity. Freshwater mussels are eaten by otters, raccoons, ducks, baboons (off the coast of South Africa) and geese, (NCRS 2007).

2.1.2. Native Marine Mussel

Mytilus sp. is a sea bivalve genus, which occurs along many coastlines around the world. *Mytilus* spp. are found throughout the world, namely in northern temperate latitudes, the Mediterranean Sea, the Pacific coast of North America, south–eastern and south–western coasts of South America, Australia, New Zealand, the Kerguelen Islands, and the Pacific coast of Asia (Gosling 1992).These mussels are widely cultured, not only in Europe, but in the Pacific coast of China, Japan and Korea as well. This mussel genus is very competitive and, in some cases, able to colonize the entire available substratum (Lachowicz 2005).

Mussel (*Mytilus* spp.) production is one of the most economically important aspects of global aquaculture. The farming of mussel (*Mytilus* spp.) is without doubt the most efficient way to convert the organic matter produced by marine organisms in the first link of the food chain (Culha et al. 2008). In Europe, production is known to consist of *M. galloprovincialis*, *M. edulis* and a third, less profitable species *M. trossulus,* but only in the Baltic Sea. The annual world production of mussels *Mytilus galloprovincialis* is 1.690.835 t (Stankovic et al. 2011).

Distribution and Habitat

Marine mussels (*Mytilus* spp.) are abundant in the middle shore to the shallow sublittoral zone in temperate seas globally and attach to suitable substrates such as piers, rocks and stones with protein threads that are the fibrous byssus. Other species of marine mussel live in tropical intertidal areas, but not in the same huge numbers as these in temperate zones. They may occur on soft sediments in estuaries, often forming large beds. Marine mussels are farmed commercially in many areas (Stankovic et al. 2011, Stankovic and Jovic 2012). Certain species of marine mussels prefer salt marshes or quiet bays, while others thrive in pounding surf, completely covering wave–washed rocks. Some species have colonized abyssal depths near hydrothermal vents. The South African white mussel exceptionally doesn't bind itself to rocks, but burrows into sandy beaches extending two tubes above the sand surface for ingestion of food and water and exhausting wastes.

They occur from the high intertidal to the shallow subtidal zone attached by fibrous byssus threads to suitable substrata. Found on the rocky shores of open coasts attached to the rock surface and in crevices, and on rocks and piers in sheltered harbors and estuaries, often occurring as dense masses. Marine mussels are usually found clumping together on wave–washed rocks, each attached to the rock by its byssus.

Mussel growth rates are known to considerably vary between latitudes. The highest growth rate of the common mussel is specific for the Atlantic coast of Spain (Vigo Bay) (Zotin and Ozernyuk 2004). It is possible that these growth differences are due to specific micro–environmental conditions in each of these biotopes, including the rate of tidal currents, temperature, oxygen content, and food supply. A temperature is the main climatic factor of marine mussel growth. In addition, the habitat of the mussel (various littoral, sublittoral, estuary regions, etc.) considerably influences their growth (Kulakovskii 2000).

M. galloprovincialis, *M. edulis* and *M. trossulus* are closely related species, the identification of which has long been controversial. More accurate species identification is now possible using genetic methods. The three species are known to hybridize wherever their geographic ranges overlap. Well–studied hybrid zones of *M. galloprovincialis* include the areas between southwest France and the Scottish coast between *M. galloprovincialis* and *M. edulis*; on the west coasts of North America between *M. trossulus, M. galloprovincialis* and *M. edulis* (Stankovic et al. 2011).

The Mediterranean mussel *M. galloprovincialis* (Lamarck) is known from classical texts of Greek antiquity and their use in diet and medicine by humans in the Mediterranean coastal areas (Stankovic et al. 2011). *M. galloprovincialis* is endemic to the Mediterranean Sea, the Adriatic Sea and the Atlantic Ocean from Ireland to Morocco. The Mediterranean mussel is a warm–water invasive species, frequently transported by human activities. From the Mediterranean, *M. galloprovincialis* as invasive species has colonized the sea waters of China, Korea, southeast Australia, Hawaii, Mexico, California, the west and east coast of Canada and arrived in South Africa (Stankovic et al. 2011). The mussel *Mytilus galloprovincialis*, a native of the Mediterranean Sea, has spread extensively, and now is a successful invader on the central and southern coasts of California, where it has largely displaced the native congener, *Mytilus trossulus* (Lockwood and Somero 2011) and it is now the most abundant alien marine species along the South African coast and shows partial habitat segregation with the indigenous mussel *Perna perna* (Rius and McQuaid 2006).

It is known that *M. galloprovincialis* is able to out compete and displace native mussels and has become the dominant mussel species in certain localities because *M. galloprovincialis* may grow faster than native mussels. Thus, *M. galloprovincialis* is both ecologically invasive and a source of 'genetic pollution' that may threaten the genetic integrity of a native mussel species over much of its geographic range (Stankovic et al. 2011).

In its native range, *M. galloprovincialis* can be found from exposed rocky outer coasts to sandy bottoms. Only the Mediterranean mussel occurs on open coasts, while all other kind of mussels is restricted to bays and estuaries. The Mediterranean mussel tends to grow up to 15 cm, although typically only 5–8 cm. A 5 cm–long mussel can filter 5 liters water/h. While mussels grow to optimum size in sea water of temperature 15°C–25°C, their biological activities are decreased over 25 °C. They are resistant to 5–40 ‰ salinity, but the optimal degree of salinity is between 20 ‰ – 35 ‰. The mean density in the most crowded beds may reach 24,000 mussels / m^2 (Stankovic et al. 2011).

People harvest Mediterranean mussels as food (Joksimovic et al. 2011, Jovic et al. 2012) and they are used in commercial aquaculture (Stankovic and Jovic 2012). These marine mussels are considered an important food source in some coastal areas and the shells are used in jewelry manufacturing. Mussels also help limit algae growth, which has become problematic in the Mediterranean Sea (MarLIN 2006). Large marine mussel beds provide habitat and prey for other animals and act as a substrate for algal attachment, increasing local diversity. Mussel larvae are an important food source for various marine animals (MarLIN 2006).

Feeding

Marine mussels are filter feeders. They feed on plankton and other microscopic sea creatures which are free–floating in seawater. A mussel draws water in through its incurrent siphon. The water is then brought into the branchial chamber by the actions of the cilia located on the gills for cilliary–mucus feeding. The wastewater exits through the recurrent siphon. The labial palps finally funnel the food into the mouth, where digestion begins (MarLIN 2006).

Mussels feed on organic matter and phytoplankton and can move with the aid of a foot. The foot of a mussel is an important organ because it enables the mussel to attach itself by byssus threads to a solid substrate. Byssal attachment is an important, sometimes critical, factor in mussel aquaculture (Stankovic et al. 2011, Stankovic and Jovic 2012).

Reproduction

Marine mussels are gonochoristic, with separate male and female individuals. In most marine mussels, fertilization occurs externally or outside the body by spawning their free–floating gametes into the water column, with a larval stage or development is free–swimming and planktonic that drifts for three weeks to six months, before settling first on filamentous organisms such as seaweeds.

The two main reproductive cycles of *M. galloprovincialis* varies greatly according to geographic region and is characterized by one or two major spawning events per year, in spring and autumn, but depending on environmental temperature and food availability. Mussels may spawn just once or several times each year. Immediately after the reproduction process, a mussel can lose up to 70 % of its soft tissue. The sexes are separate in mussels, but there are no morphological differences between males and females. Factors such as water temperature, nutrient availability and the reproduction cycle of mussels can influence meat yield and biochemical composition of these mussels (Stankovic et al. 2011, Stankovic and Jovic 2012).

2.2. Habits of Nonnative Mussels

Species distribution changes naturally over time. Human activities greatly increase the rate and the spatial scale of these changes by accidentally or deliberately moving organisms across the world. Besides the marine mussel (*M. galloprovincialis)*, also freshwater mussels have a high potential for rapid adapting to new environmental conditions, which could lead to significant long-term impacts in the waters they inhabit (Mills et al. 1996).

The review of Russian, European, and North American literature (Ludyanskiy et al. 1993) reveal the optimal environmental requirements of the zebra mussel. Its primary requirements are salinity of 0–4 ppt, average summer water temperature of 17–23 °C, pH level of 7.4–9.0, calcium concentration of 20–125 ppm, and dissolved oxygen content of 8–10 ppm. Its secondary requirements are water velocity of 0.2–1.2 m/sec and the presence of solid substrates (rocks, wood, or gravel). The research in North America has also demonstrated that zebra mussels are capable of living in brackish water or estuaries where the salinity does not exceed 8 to 12 ppt, (Mackie and Schloesser 1996).

Environments with parameters close to the mentioned above will be probably colonized first, but in Russia and Europe zebra mussels do inhabit bodies of water with less than optimal ecological parameters. Because of their strong genetic plasticity, zebra mussels can tolerate changes over time in their environment to levels outside these ranges (Ludyanskiy et al. 1993). The settlement of *D. polymorpha* is remarkably similar to that has been described in *M. edulis* and other marine bivalves (Ackerman et al. 1994).

Since the discovery of quagga mussels (*D. bugensis*) in the Lower Colorado River reservoirs in the early 2007, public agencies have been scrambling to determine how to control their population and distribution. The quagga mussel (*D. bugensis*), originally from the Ukraine, infested the Great Lakes region in the 1989, competing with their earlier arrived relative, the zebra mussel (*D. polymorpha*). The quagga mussel is currently of major concern in the United States as an invasive species (Jones and Ricciardi 2005).

Quagga mussels are able to adapt to varying environmental conditions by being able to colonize hard and soft substrata in varying water depths (Mills et al. 1996). They have been found in deep water and they do not like light. In slightly deeper water where sunlight is more diffuse, *Dreissena* population densities may reach 100,000 individuals per m^2. Quagga mussels are reported to be less tolerant of high water temperatures than zebra mussels. The distribution of quagga mussels is also partially dictated by calcium concentration. Concentrations above 12 mg/l are needed to form the shell material (Jones and Ricciardi 2005).

Besides temperature, another important environmental variable that affects quagga and zebra mussel distributions and eventual population densities is calcium concentration. Mussels require calcium for basic metabolic function and for shell growth. Based on field distributions, quagga mussels apparently have a slightly higher calcium requirement than zebra mussels. Guagga mussels were not found in waters with calcium concentrations lower than 12 mg/L, while zebra mussels were present (but not abundant) at concentrations as low as 8 mg/L (Jones and Ricciardi 2005). High calcium concentrations (>80 mg/L) may favor quagga mussels over zebra mussels (Zhulidov et al. 2004). A summary of mussel tolerance limits for other environmental variables, such as dissolved oxygen, pH, and salinity, is provided by Cohen (2007). Information about the distribution of both species in North America and Eurasia provides evidence that *D. polymorpha* and *D. bugensis* may have different tolerances to salinity and temperature. *D. polymorpha* is more tolerant in increases of salinity than is *D. bugensis* (Miils et al. 1996). Both, zebra and quagga mussels can survive cold waters near freezing, but cannot tolerate freezing. Quaggas need waters above 9°C in order to reproduce and generally prefer colder waters to survive and they are more efficient filter feeders than zebra mussels. Quagga mussels are typically found at any depth as long as oxygen is present. Both species prefer to avoid light and are usually found in shaded areas or below the depth that light penetrates water (USFWS 2007). *Dreissena* mussels are efficient filter feeders that can siphon more than one liter of water per day by using cilia to pull water, phytoplankton, zooplankton, bacteria, algae, organic debris, even their own larvae, silt, and clay, through an inhalant siphon (MacIssac 1996). The golden mussel, *Limnoperna fortunei* (Mytilidae) was introduced into Argentina from Asia in the year 1991 and within a decade spread to four other South American countries. The invasion of the zebra mussel (a morphologically similar bivalve) into North American waters demonstrated the vulnerability of industrial water conduits to macrofouling species.

Table 1. Tolerance limits of *L. fortunei* compared to those of *D. polymorpha* (Ricciardi 1998)

	L. fortunei	*D. polymorpha*
Salinity	0-12‰	0-5‰
Calcium	≥3.0 mg/	≥12.0 mg/L
pH	≥6.4	≥7.4
Temperature		
Larval Development	16-28°C	12-24°C
Adult Survival	8-35°C	2-30°C
Oxygen	≥1.0 mg/L	≥1.8 mg/L

Although these two species share some of the physical characteristics, the golden mussel exhibits a wider tolerance of ecological parameters than the zebra mussel, (Ricciardi 1998), impacts may be similar to *Dreissena*. Therefore, if introduced into North American waters, the golden mussel is expected to invade a broader range of habitats.

L. fortunei is primarily a freshwater species, capable to tolerate brackish waters and maintaining substantial populations in estuarine habitats. Their shells appear golden or yellowish in color. They are filter feeders on phytoplankton and zooplankton and they are food source for fish. The golden mussel is tolerant of polluted and contaminated water conditions, capable of inhabiting waters with relatively low calcium and pH, Table 1. Attach byssally to available substrates, forming dense aggregations, often establishing colonies with densities approximately 150,000 ind/m^2 (Darrigran 2002) and females typically comprise two-thirds of population. Spawning occurs 1-2 times per year and external fertilization produces free-swimming veligers.

These invasive mussels have a high potential to adapt rapidly in new environmental conditions, which could lead to significant long–term impacts in the waters they inhabit. Local increases in sedimentation are a common consequence of aggregations of invasive mussels that can have important effects on other organisms (Sousa et al. 2009).

3. Mussels Ecology

Ecological Values of Mussels

Freshwater mussels typically inhabit unpolluted, standing or flowing waters that are rich in oxygen, calcium, and suspended food particles (Helfrich et al. 2009). Mussels are an important part of aquatic ecosystems: filtering out nitrogen and bacteria they improve water quality. Despite the biofiltering value freshwater mussels provide, their population is decreasing rapidly. During the past 30 years numbers of both, individuals and species diversity of native mussels, have declined in the United States and Canada (Strayer et al. 2004). In Britain, the freshwater pearl mussel is only found in viable populations in some Scottish rivers. Other populations do not have active recruitment and could soon become extinct (Skinner et al. 2003). Reduction in aged mussel numbers can be a result of pearl fishing, but higher temperatures, eutrophication and increases in nitrate levels may also reduce longevity. Most mussel species require a slightly higher than average amount of oxygen in the water in order to live. It is widely accepted that eutrophication is a major cause of mussel decline, and agricultural practices, particularly efficient field drainage and the use of fertilizers are major contributors to this process (Skinner et al. 2003).

Strong evidence indicated that some form of habitat destruction and loss have led to decline and extinction of freshwater mussels populations. Dam construction is one of the most common causes of habitat loss. Habitat modification through man–made structures like dams and channel alterations has destroyed free–flowing water habitats. These modifications restrict many mussels from dispersing which results in small, isolated populations. In addition to habitat modifications, mussel populations are exposed to toxic runoff containing fertilizers, herbicides and pesticides from land use practices (Schaffner et al. 2009). According to Downing et al. (2010) specific impacts have derived from pollution and water quality degradation and disturbance by human recreation. Such factors, direct exploitation, as well as exotic and invasive species and global change have captured the public interest.

Freshwater mussels are an important part of the food web in aquatic ecosystems. Adults are filter–feeders and consume phytoplankton, diatoms, and other microorganisms in the water column, as well as detritus and bacteria. Mussels are, in turn, consumed by muskrats, otters, and raccoons and young mussels are often eaten by ducks, herons, and fishes, as well as other invertebrates. As juveniles, mussels are deposit feeders and use their ciliated foot to obtain nutrients. Effects of mussels on nutrient cycling and translocation vary with hydrologic and trophic conditions, but conversion of suspended material to dissolved nutrients by mussels stimulates primary production where nutrients are limiting (Vaughn et al. 2007). The Unionoida genera are recognized to be among the most sensitive of families tested to specific contaminants (Mosher 2008): they have shown a higher sensitivity to ammonia than other organisms.

Mussels can act as habitat modifiers because of the way they take up food. As natural filter feeders, mussels strain out suspended particles and pollutants from the water column and help improve water quality. Some mussels can filter up to 50 liters of water per day, which helps to improve water quality for other animals, including humans. In the same time filtering processes create a possibility of removing chemical compounds, such as organic compounds, heavy metals and nutrients, from water (Brzozowska et al. 2012). Freshwater mussels in the tidal Hudson River in New York filtered nearly 5.3 million gallons of water per day, approximately equal to the daily freshwater discharge of the Hudson River during the summer (Strayer et al. 2004).

Mussels are commonly used as indicators or biological monitors of present water quality conditions in rivers, lakes and seas (Sures 2004, NRCS 2007, Stankovic et al. 2011, Stankovic and Jovic 2012). If, for example, mussels change their shell–opening frequency, these alterations can indicate the presence of toxins within the water. This shell–opening behavior can be monitored online using the commercially available mussel monitor with *D. polymorpha* or *M. edulis* as indicator species for monitoring contamination in freshwater or marine habitats, respectively (Sures 2004). A sudden increase in mortality of freshwater mussels is a reliable indicator of toxic contamination or the disappearance of freshwater mussels usually indicates chronic water pollution problems. Moreover, biologists can measure the amount of pollutants found in mussel shells and tissue to determine the type, extent, and even timing of water pollution events in rivers, lakes and sea coasts (NRCS 2007, Stankovic et al. 2011).

The amount of pollutants can be measured in mussel tissues to determine the type and extent of water pollution (Bervoets et al. 2005, Mosher 2008, Stankovic et al. 2011, Anzano et al. 2011, Jovic et al. 2012, Stankovic and Jovic 2012, Markovic et al. 2012). Also, biomedical uses of mussels are presently being studied. Evidence from cancer research suggests that some mussels may be resistant to certain types of cancer and the extraction of cancer–curing drugs from mollusks may be feasible in the future (Helfrich et al. 2009).

Although freshwater mussels have little value as human food, they are a major food item for valuable wildlife such as muskrat, otter, and raccoon. Young mussels are eaten by ducks, herons, and sport fish. As important natural filterers in the riverbed, they improve water quality by straining out suspended particles and pollutants from rivers and lakes, and they help to remove and store contaminants, making the water more fit for human uses. Because of their filtering capacity, mussels are an integral part of the natural purification process in rivers and lakes. They are particularly useful in removing algae and suspended particles from turbid and organically enriched waters near wastewater facilities and in fish farm effluents (Helfrich et al. 2009).

Live mussels and dead shells also provide habitat for a variety of aquatic insects and algae, providing the foundation for a variety of life forms and habitat conditions suitable for other aquatic organisms. Living mussels and their spent shells provide or improve habitat for other organisms through: 1) the physical structure of the shells, 2) sediment stabilization and bioturbation, and 3) biodeposition of organic matter and nutrient excretion, which directly and indirectly influence food availability. Spent shells are important as habitat in areas where other hard substrata are unavailable (Gutierrez et al. 2003). However, the importance of mussels as habitat generators extends beyond provision of physical structure and often is consequent to the activities of living mussels. For example, macro–invertebrate densities are higher in mussel beds (aggregations of mussels) than outside beds, and macro–invertebrates aggregate on sediments with mussel biodeposits (Vaughn et al. 2008).

Invasive mussels are habitat generators. By reviewing the ecology of the invasive mussels, there are numerous mechanisms via which they modify, maintain and/or create habitats. Usually their high densities and broad spatial distributions, activities of such mussels can significantly alter ecosystem structure and functioning, e.g., changes in sediment chemistry and grain size, organic matter content, increased light penetration into the water column due to filter feeding.

For example *Dreissena* species feed so efficiently on the total phytoplankton community (Barbiero et al. 2006) and reduction of phytoplankton biomass leads to water clarification and sunlight penetration deeper into the water column. Increased sunlight penetration increases the heat potential and encourages aquatic plant growth at deeper levels in the aquatic environment, modifying the local ecosystem. Less phytoplankton in the water column also affects the growth potential of larval fish as their food sources diminish, resulting in decreasing sizes in adult fish populations.

In addition, changes in ecosystem structure and functioning due to engineering by invasive bivalves often have very large economic impacts. The varied ways in which they physically modify habitats, their engineering effects, should receive more serious consideration (Sousa et al. 2009).

Dreissena have also been shown to selectively reject toxic cyanobacteria through pseudofeces in favor of non–toxic species such as green algae and diatoms, enriching the toxic species along the lake bottom and during cyanobacteria bloom events (Barbiero et al. 2006, Juhel et al. 2006). Microcystins, a cyclic peptide produced by *Microcystis*, a colonial cyanobacterium, pose significant ecological problems and health issues to plants and animals, including humans (Vanderploeg et al. 2001). Even if not directly consumed by *Dreissena*, some invertebrate organisms experience a major population decline as they are either out–competed by the mussels for the phytoplankton food source (Wilson 2003).

Ecologically, the zebra mussels cause more problems. One such problem is that they need to attach to a hard surface to survive, and these hard surfaces could be anything from many manmade objects up to other animals. Zebra mussels will attach to crayfish, turtle shells, as well as other mussels. When a native mussel has zebra mussels attached, the native mussel loses its ability to move, feed, breath, and breed. Eventually this will lead to the death of the native mussel. The golden mussel has the potential to affect the diversity of native molluscan communities e.g., can overgrow and possibly kill native mollusks; the impact is similar to the impact caused by the zebra mussel in the Northern Hemisphere (Darrigran 2002).

On the other hand zebra mussels have an important ecological role in aquatic habitats by filtering suspended particles they clarify the water, and may generally improve water quality, and used in water quality management, (Mackie and Schloesser 1996). In North America, zebra mussels serve as an excellent biological indicator of chemical, viral, and bacterial pollutants in the Great Lakes and the St. Lawrence River, (Graczyk et al. 2008), because they can bioaccumulate these pollutants in their tissue. As the mussels grow in two years, they fully cover the structures and filter the water passing through them removing nutrients, bacteria, and other suspended organic substances. It is well known that trace metals can accumulate from the dissolved phase via absorption across the body surfaces and another route is absorption through the gastrointestinal tract from food, e.g., algae, bacteria, sediment particles. According to presented work of Brzozowska et al. (2012) it seems that substances were also removed from seawater as a result of bivalves' filtering activity and adsorption onto the shell and body surfaces. Smaller individuals have a higher filtration rate and are able to take up metals or other substances more quickly (Brzozowska et al. 2012).

MacIsaac (1996) claimed that populations of the zebra mussel can remove 90 % of organic matter from water. In the Brzozowska et al. (2012) experiment *Mytilus sp.* removed less organic matter and the maximum observed removal efficiency was 47 %. Elliott et al. (2008) suggested using *D. polymorpha* as a kind of an on–site biofilter for freshwater treatment from algal biomass. Schernewski et al. (2012) focused on zebra mussels farming because they are efficient in removing nutrients and improving water transparency in the Oder Lagoon. They found that mussel farming could potentially remove nearly 1000 t of N (70 t of P)/year, or about 2 % of the present N and P loads, and it would have the additional benefit of improving water transparency. Similarly to *D. polymorpha*, *Mytilus* sp. is very competitive (Lachowicz 2005).

For example, *Mytilus* spp. is able to live in installations of seawater desalination plants. The current investigations have revealed that it is possible to improve seawater quality and attain high efficiency of removal of some heavy metals and – to a lesser degree – organic substances from raw seawater using the filtering activity of *Mytilus* spp. (Brzozowska et al. 2012). These results suggest that it might be possible to use also *Mytilus* sp. as a kind of biofilter for the seawater quality improvement.

Mussels are particularly important in transferring nutrients from the water column to the sediment. Mussels provide nutrients to other organisms and to the mussel community through excretion of NH_3 and P, through biodeposition of feces and pseudofeces, and by release of stored nutrients as dead mussels decay (Vaughn et al. 2008). That increases habitat complexity or heterogeneity, causing species abundances. Accordingly, many invasive mussels enhance organism's density, biomass and/or diversity both in soft sediments and rocky bottoms of marine and freshwater ecosystems (Sousa et al. 2009).

The arrival of *M. galloprovincialis* to the marine coasts of South Africa increased benthic invertebrate density and species richness. Increase in the density of benthic invertebrates and biomass were also observed in association with the golden mussel *L. fortunei* presence in South American rivers with regard to areas without them (Sousa et al. 2009). Some studies (NRCS 2007) have shown a negative impact of *D. polymorpha* beds on the diversity and density of invertebrate.

4. Mussels Control

4.1. Pollutants

The increasing worldwide pollution of streams, rivers, lakes, seas and oceans with thousands of industrial and natural chemical compounds is one of the key environmental problems facing humanity. Today, contamination of the environment is apparent not only in areas of direct anthropogenic influence, but also in remote areas as a result of long range atmospheric transport and deposition of pollutants. Agricultural activities results in the release of organic matter and nutrients into the environment which can contribute to the eutrophication and cause major changes to aquatic community dynamics. Many chemicals used in farming, plant protection products, biocides (pesticides, herbicides, insecticides and fungicides), veterinary drugs, may contaminate the aquatic environment (Schaffner et al. 2009). Human habitations can also be major sources of organic matter release into aquatic environments, particularly in coastal areas, via wastewater release (Heinzow et al. 2007, Schaffner et al. 2009). Inland and coastal surface waters can be also contaminated by human waterborne zoonotic enteropathogens originating predominantly from wastewater treatment plant effluents and agricultural runoff (Graczyk et al. 2008). Incinerators and domestic heating equipments release persistent organic pollutants (POPs), such as dioxins, PCBs, PAHs, etc. (Lewtas 2007, Van Caneghem et al. 2010). Industrial activities also release a very wide range of toxic chemicals. Transport and energy production release such substances as PAHs, trace elements, radionuclide and many atmospheric pollutants (England et al. 2001). Persistent inorganic and organic contaminants as well as waterborne pathogens in aquatic systems are a global problem, since continuous exposure of aquatic organisms to their low concentrations may result in bioaccumulation and subsequent transfer to man through the food web.

4.2. Bioindicators

Bioindicators include biological processes, species or communities and are used to assess the quality of the environment and how it changes over time. Changes in the environment are often attributed to anthropogenic disturbances (e.g., pollution, changes in land use such as agricultural intensification) or natural stressors (e.g. climatic changes, drought, late spring freeze), although anthropogenic stressors form the primary focus on bioindicators research (Holt and Miller 2011).

Worldwide mussels have been used as bioindicators of water quality i.e., to assess concentrations of bioavailable contaminants in waters. Mussels, as filter–feeding animals, through feeding assimilate not only the food necessary for growth and development, but also contaminants present in the water. They process large volumes of water thereby accumulating polluting compounds in concentrations several orders of magnitude greater than those in the ambient water. Several attributes that make mussels superior than other organisms for environmental monitoring include the next: wide geographical distributing, abundant, sedentary, tolerant to environmental alterations, tolerant to various environmental contaminants, high bioconcentration factors of pollutants, very low–level metabolizing enzyme activities of organic contaminants, wide and stable populations, reasonably long–lived, reasonable size, sturdy enough to survive in laboratory and field studies in cages (Zhou et al. 2008).

Marine mussels are recognized as pollution bioindicators because they accumulate pollutants in their tissues at elevated levels in relation to pollutants in a marine environment (Potrykus et al. 2003, Jovic et al. 2011). For this purpose different varieties of mussels were used, mostly species within the family Mytilidae, genus *Mytilus*. Freshwater mussels are used less as bioindicators. It appears that freshwater species may be less sensitive than their marine relatives to pollutants, in general, although there are insufficient freshwater data to be sure if there is a real difference in their sensitivity. The zebra mussel is the most highly applicable bioindicator dealing with antropogenic impacts on freshwater environments (Griebeler and Seitz 2007). It has been established as a bioindicator for passive as well as active biomonitoring of freshwater ecosystems (Bervoets et al. 2005, Pain et al. 2005, Anzano et al. 2011). The pearl mussel, family *Margaritiferidae*, also have been recommended as good biomonitors for studying seasonal, annual, past and present freshwater quality conditions and exposures to various pollutants (Webb et al. 2008).

Pollutants almost always follow the cyclic order: industry, atmosphere, soil, water, fish, and humans, and they have the ability to bioconcentrate in organisms directly from the water, bioaccumulate and biomagnify in the food chain, causing higher trophic organisms to become contaminated with high concentrations of chemical contaminants (Stankovic and Stankovic 2012). The various toxins found in waters can be grouped into three main categories: inorganic pollutants – heavy metals, organic pollutants and pathogenic bacteria. Number of researchers demonstrated the reliability of mussels as bioindicators of mentioned pollutants.

4.2.1. Inorganic Pollutants – Heavy Metals

Inorganic chemical pollutants are naturally found in the environment but due to human development these pollutants are often concentrated and released into an environment where cause contamination of water, air and soil. The primary inorganic pollutants of concern are heavy metals and nutrients.

Heavy metals are introduced in aquatic systems as a result of soils and rocks weathering, from volcanic eruptions and from a variety of human activities involving the mining, processing or use of metals and/or substances that contain metal pollutants (Jain 2004). Heavy metals fall into one of two categories, essential and non–essential. All heavy metals are potentially toxic at some concentration, but non–essential are particularly toxic at relatively low doses, (Stankovic et al. 2011, Stankovic and Stankovic 2012). Essential metals are required for the optimal functioning of biological and biochemical processes, and non–essential elements, such as Cd, Pb, Hg and As, have no known biological function in organisms and exert their toxicity by competing with essential metals for active enzyme or membrane protein sites (Stankovic and Stankovic 2012).

As filter–feeding organisms, freshwater and marine mussels are exposed to heavy metals that are dissolved in water and in interstice water of sediments, associated with suspended particles and deposited in bottom sediments. Thus, both freshwater and marine mussels can bioaccumulate certain metals to concentrations that greatly exceed those dissolved in water and which can produce deleterious effects on organisms. Accumulated heavy metals in mussel tissues are proportionally higher to the degree of environmental contamination. The relatively small increase of heavy metal concentration due to the ambient pollution will be reflected in measurable amounts in their tissues (Fang et al. 2003).

Marine musssels are used as bioindicators of heavy metal pollution since the year 1976, the so–called Mussel Watch program initiated in the United States of America. In the mussel watch program four marine bivalve species (*Mytilus edulis*, *M. californicus*, *Crassostrea virginica* and *Ostrea equestris*) at more than 100 sample sites were used to monitor metal pollution in the USA. Since then bivalves, notably mussels have been used to assess the levels of heavy metal pollutions in marine ecosystems worldwide. Mussels (*Mytilus edulis, M. galloprovincialis* and *M. californicus*), oysters (*Crassostrea virginica* and *Ostrea edulis*) and clams (*Mercenaria mercenaria*, *Venerupis* spec. and *Macoma balthica*) are the most commonly used bivalve groups for toxic metals bioindicators (Oehlmann and Schulte–Oehlmann 2003). The mussel *Perna virdis* is suitable bioindicator for heavy metal pollution in Asian countries, Malaysia (Kamaruzzaman et al. 2011), Hong Kong (Liu and Kueh 2005), India (Sasikumar et al. 2006), and the mussel *Mytella strigata* for the same purpose proposed in central and south America (Osuna–Lopez et al. 2009). It was also confirmed that the mussel *Perna perna* is efficient toxic metals bioaccumulator and was employed as an indicator of heavy metal pollution in Asia–Pacific, Brazil and in tropical areas (Bellotto and Miekeley 2007).

At the same time, the potential of freshwater bivalves as heavy metals bioindicators was also investigated (Oehlmann and Schulte–Oehlmann 2003). One of the most frequently used freshwater bivalve accumulator species is the zebra mussel *Dreissena polymorpha.* Zebra mussel, a widespread invasive species in Europe and USA, accumulates high amounts of heavy metals and therefore considered as an important bioindicator of freshwater environments reflecting the site specific pollution (Anzano et al. 2011, Rahnama et al. 2011, Minguez et al. 2012). Widespread from the Arctic and temperate regions of western Russia through Europe to the northeastern seaboard of North America, the freshwater pearl mussel *Margaritifera* spp. has been also recognized as good bioindicator for metal pollution (Webb et al. 2008).

The investigated metals were wide ranging, but the majority of studies, including both freshwater and marine mussels, focused on Zn, Cu, Pb and Cd (Stankovic and Stankovic 2012), and less frequently on Hg, Ni, Al, Fe, Mn, As and Cr, whilst a very sparse amount of data exists for Co, Mo, Sb, Li, Na, Ca (Goodyear and McNeill 1999).

4.2.2. Organic Pollutants

In recent years particular attention has been paid to the occurrence and role of various persistent organic pollutants (POPs) in aquatic environmental contamination. POPs are organic (carbon–based) chemicals that remain in the environment for long period of time (WHO 2010). POPs are released into air, water and land and then enter the food chain. They are globally distributed through the air and ocean currents, travelling long distances via air–water exchange and cycles involving rain, snow and dry particles. They are also transported by polluted living organisms. They are thus found in places far away from industrial sites or from agricultural areas where they were released.

POPs originate mainly from uses in industrial processes, waste incineration and agriculture. They include a wide variety of chemicals, from petroleum compounds and their derivatives (including the polycyclic aromatic hydrocarbons, PAHs) to organohalogenated compounds, such as the chlorinated pesticides DDT, aldrin and dieldrin; the polyhalogenated biphenyls (e.g. PCBs and PBBs); fire retardants such as the polybrominated diphenyl ethers (PBDEs); and most recently, the per– and poly–fluorinated compounds which have been widely used in surface applications on a variety of household and consumer items (Wua et al. 2008). They bioaccumulate and biomagnify as they move through the food chain (Norstrom 2002). For most people, exposure to POPs comes from the ingestion of food such as fish, meat and dairy products although exposure can also occur through inhalation and dermal absorption. In humans and animals, there are known adverse health effects of exposure to high levels of POPs. The effects may include cancer, damage to the nervous system, reproductive disorders or disruption of the immune system (WHO 2010).

Due to growing concerns regarding persistent organic pollutants in the environment, extensive studies and monitoring programs have been carried out in the last two decades to determine their concentrations in water, sediment and more recently, in biota. Mussels have been used extensively as sentinel indicator species for monitoring uptake and accumulation of POPs, since they exhibit high potential for biomonitoring and represent one of the important food products for human consumption.

From the year 1979 the Reseau National d'Observation de la qualité du milieu marin (RNO) has been collecting mussels and oysters and analyzing POPs from about 90 sites along the French Coast (Beliaeff et al. 1998). In the year 1986 the NOAA National Status and Trends (NSandT) Program Mussel Watch has monitored concentrations of persistent organic pollutants in the coastal areas of United States. In the year 1992 the United Nations Intergovernmental Oceanographic Commission (UN/IOC) inaugurated the International Mussel Watch Program by collecting and analyzing mollusks from 76 sites in Latin America (Farrington et al. 1995, Sericano et al. 1995). This was followed by a program in Asia (India, Philippines and Thailand) in 1994–1997 (Tanabe et al. 1998, Tanabe 2000). In the year 1996 the United Nations Development Program (UNDP) and World Bank jointly sponsored samples collections and analysis in the Black Sea (Moore et al. 1998). Also a number of authors were measured various POPs (PAHs, PCBs, DDT, TBT etc.) in different varieties of *Mytilus spp.*: *M. trossulus* (Potrykus et al. 2003, Page et al. 2005); in blue mussels *M. edulis* (Chou et al. 2003, Tronczynski et al. 2004, Ruus et al. 2006, Skarpheoinsdottir et al. 2007, Devier et al. 2005); in *M. californianus* (Oros and Ross 2005); in Mediterranean mussels *M. galloprovincialis* (Carro et al. 2005, Namiesnik et al. 2008), in zebra mussels *D. polymorpha* (Binelli et al. 2001, Binelli and Provini 2003, Kwon et al. 2006, Minier et al. 2006) and Quagga mussels *D. bugensis* (Turner et al. 2011). The median concentrations of POPs among all these areas differ to some extent. POPs are generally higher in the industrialized nations and for some POPs there are good correlations between concentrations in mussels and numbers of people living near a collection site.

The international community has called for actions to reduce and eliminate production, use and releases of these substances. To the end, two international legally binding instruments have been negotiated and concluded: the Protocol to the regional UNECE Convention on Long–Range Transboundary Air Pollution (CLRTAP) on POPs, entered into force in the year 2003, and the global Stockholm Convention on POPs, entered into force in the year 2004. These instruments establish strict international regimes for initial lists of POPs (16 in the UNECE Protocol and 12 in the Stockholm Convention). Both instruments also contain provisions for including additional chemicals into these lists. They lay down the control measures such as prohibition or severe restriction of the production and use of intentionally produced POPs, restrictions on export and import of the intentionally produced POPs, provisions on the safe handling of stockpiles, provisions on the environmentally sound disposal of wastes containing POPs, provisions on the reduction of emissions of unintentionally produced POPs (e.g. dioxins and furans).

In terms of protecting the environment from further harmful effects of POPs, a large number of directives have been prescribed and adopted. Directive 2001/201/EC, Directives 2001/2375/EC, Directive 2002/69/EC, Directive 2003/806/EC and Directive 1986/363/EC set limit values for POPs and Dioxins and Furans (PCDD/Fs) in food and feeding stuffs and contain provisions for monitoring and control. Commission Regulation 2006/199/EC sets limit values for PCDD/Fs and dioxin like PCBs for meat and meat products, fish and fishery products, milk and milk products, eggs, oils, and fats. The limit values are based on the tolerable weekly intake of 14 pg WHO TEQ/kg body weight for the sum of PCDD/Fs and dioxin like PCBs.

4.2.3. Pathogenic Bacteria

An organism is pathogenic if it causes disease in its host. Some species of pathogenic bacteria kill their host cells, but the largest number produce toxins that cause damage to the host cell metabolism. Some pathogenic bacteria can be transmitted to humans via food. Bacterial food–borne pathogens may be grouped into those that cause food intoxication and those that can result in food–borne bacterial infection.

Mussels are able to retain and to concentrate *Escherichia coli*, the fecal contamination indicators (Selegean et al. 2001, Normanno et al. 2006). Since bivalves do not normally contain fecal coliform bacteria in their gut, the presence of such bacteria in a mussel indicate on a recent environmental exposure of the mussel to bacteria and making them a good bioindicator of *E. coli* in the environment under conditions that conventional monitoring may often miss. *E. coli* food infection causes abdominal cramping, water or bloody diarrhea, fever, nausea and vomiting. Bacteria usually have a short residence time in the mussel and usually can be removed by a process of depuration. Bacterial indicators for the classification of production areas, categories A, B or C, are currently established by the evaluation of *E. coli* in the mussels flesh and fluid. Category A corresponds to the lower concentration of 230 colony forming units (cfu) of *E. coli* in 100 g mussel flesh. These mussels can be sold for direct human consumption. Categories B and C imply lower concentration of 4600 cfu of *E. coli* in 100 g mussel flesh and cannot be sold direct for human consumption. These must be purified, relayed or heat treated prior to the sale, Council Regulation (EC) 854/2004.

Bacteria such as members of the genera *Salmonella*, *Vibrio* and *Aeromonas* are also frequently present in mussels. Some species of *Salmonella* spp., *Vibrio* spp. and *Aeromonas* spp. are recognised as human pathogens causing enteric pathologies, wound infection and septicemia (Gugliandolo et al. 2010).

Normanno et al. (2006) report that in 600 samples of *M. galloprovincialis* collected from retail outlets in the Puglia region, Italy, *Vibrio* spp. were found in about 10 % of mussels. One sample was contaminated with *Salmonella* spp. but no relationship was observed between vibrios and fecal coliforms and *E. coli*. Over a two–year period in samples of mussels collected from two coastal areas of the Adriatic Sea, 726 bacterial strains were isolated. Study showed that 46.9 % of total bacteria number belonged to the *Vibrio* genus, 29.0 % was of the *Aeromonas* genus and the remaining 23.3 % was made up of the *Pseudomonas*, *Flavobacterium*, *Pasteurella*, *Agrobacterium* and *Ochrobacterium* genera. The *Vibrio* genus, isolated more often than the other genera, was particularly prevalent in summer, while it was scarce in the winter months (Croci et al. 2001). In 885 samples of blue mussels *Mytilus edulis* from 102 production sites and 5 wild–growing mussel sites along Norwegian coast *Vibrio* spp. was isolated at 11.4 %. However these results indicate that potentially pathogenic *Vibrio* spp. do not pose a considerable hazard to blue mussel consumers in Norway, but that this might change during prolonged periods of high water temperatures (Bauer et al. 2006).

4.3. Invasive Mussel Control

In the past 40 years, the rate of and risk associated with biotic invaders have increased enormously because of human population growth, rapid movement of people, and alteration of the environment (Pimentel et al. 2005). Non-native invasive species cause major environmental damages and substantial economic losses (Mehta et al. 2007). Many human activities, such as agriculture, aquaculture, recreation and transportation promote both the intentional and accidental spread of species across their natural dispersal barriers (Darrigran 2002). As trade and travel increase worldwide, the number of non-native invasive species introductions continues to rise. Not all non-native or non-indigenous species become 'invasive'. Some fail to thrive in their new environment and die off naturally. Other species survive, but without destroying or replacing native species (Lovell and Stone 2005). Some invasive species, such as the zebra mussel, have also provided benefits to the habitats they 'invade'.

Zebra mussels are one of the best studied and well-known aquatic invasive species. Ecological risks and economical impacts of zebra mussels (*Dreissena polymorpha*) include alterations in the transfer of energy and cycling of materials in aquatic systems, increased accumulation of contaminants in aquatic food chains, clogging of water intakes, and damage to related infrastructure (Wu et al. 2010). Zebra mussels colonize docks, locks, ship hulls, water intake pipes, and other mollusks, causing great damage to power plants and water treatment facilities. Controls include biocides, chlorine, thermal treatment and mechanical/manual removal. There are many estimated costs for preventing, controlling, and studying zebra mussels (Lovell and Stone 2005). A number of publications have reported that the costs of the mussel to be around $5 billions (Lovell and Stone 2005, Pimentel et al. 2005).

The zebra mussel has spread into most of the aquatic ecosystems in the eastern United States and is expected to invade most freshwater habitats throughout the nation (Lovell and Stone 2005). Another related mussel species, the quagga mussel, is spreading rapidly and in some cases displacing the zebra mussel (Pimentel et al. 2005). Large mussel populations reduce food and oxygen for native fauna. In addition, zebra mussels have been observed completely covering native mussels, clams, and snails, thereby further threatening their survival. Mussel densities have been recorded as high as 700,000/m^2. Zebra and quagga mussels also invade and clog water intake pipes, water filtration, and electric generating plants; it is estimated that they cause $1 billion/year in damages and associated control costs per year (Pimentel et al. 2005). Although some researchers reported specific total economic damages and associated control costs, precise economic costs associated with some of the most ecologically damaging exotic species are not available (Lovell and Stone 2005).

For the zebra mussel, several geographic- and time-dependent environmental factors, such as temperature, dissolved oxygen, salinity, food availability, predation pressure and environmental perturbation, have been reported to affect distribution, abundance and reproductive activity. Similar to the zebra mussel, the golden mussel (*L. fortunei)* has rapidly become a major nuisance for industrial and power plants that use raw river or lake water for their processes (Boltovskoy et al. 2009). Both species are sessile, byssate bivalves with a planktonic larval stage and extremely high reproductive capacity. For both species adults attain much higher biomass in water bodies they invade than all of the native invertebrates combined. They create substrate complexity otherwise not found in freshwater systems. Both are very active suspension feeders and act as effective ecosystem engineers. Although taxonomically unrelated, their ecosystem impacts are surprisingly similar. The golden mussel has broader environmental tolerances, Table 1, and therefore may be a much more successful invader than the zebra mussel.

Although to date the zebra mussel is considered the most aggressive freshwater invader, soon many water bodies may receive another, even more aggressive invader (Boltovskoy et al. 2009). The combination of early sexual maturity, high fecundity, and wide environmental tolerance allows *L. fortunei* to be a successful invader into the new environment. *L. fortunei* is growing attached to solid non-living substrates impacting the sources of water supply for many water-treatment plants, industrial refrigeration systems and power stations (Darrigran 2002). The impact of the golden mussel is not only restricted to economic aspects. Among the potential impacts associated with the presence of this invasive bivalve, the rapid change produced in benthic communities should be noted. The displacement of native mollusk species was recorded. On the other hand, the large biomass, associated with high densities of *L. fortunei*, may impact aquatic food chains (Boltovskoy et al. 2009).

Suggested techniques for controlling biofouling by mussels range from mechanical removal, use of sieves, manipulation of water temperature and flow rate, application of electrical currents and antifouling coatings, to the injection of biocidal agents (Boltovskoy et al. 2009). The genus *Dreissena* which produces millions of larvae has a high potential for rapid adaptation (Mills et al. 1996). The factors that will limit the distribution and abundance of zebra mussels appear to be calcium content, temperature, and pH (Mackie and Schloesser 1996). In the case of *L. fortunei*, in addition to temperature, two major factors may regulate its reproductive activity: the food availability and blooms of toxic cyanobacteria significantly shortening the otherwise very long reproductive period. This information is important for the design of antifouling programs, involving the use of molluscicides, has potential for reduced biocide use (Boltovskoy et al. 2009).

There are few theoretical and even fewer empirical studies dealing with the economic costs of invasive mussel species control (Pimentel et al. 2005, Lovell and Stone 2005). The literature is still in its infancy. The biological and economic data can be combined to determine the least cost strategy for control at various life stages of the invasive mussel. In general, species that are short lived with high reproductive rates are more effectively controlled by reducing eggs and juveniles, but when adults are long-lived and with low reproduction, it is more cost effective to target adults (Lovell and Stone 2005). The aquatic studies obtaining cost estimates, reviewed in the paper of Lovell and Stone (2005), show values ranging from several hundreds of thousands of dollars a year to tens of millions of dollars a year. The most obvious point of the paper is that the literature is still in its infancy.

The increasing economic and environmental losses caused by non-native invasive mussels amplify the value of identifying and implementing optimal management options to prevent, detect, and control the invasive mussels. Previous literature has focused largely on preventing introductions of invasive mussels and post-detection control activities, but only a few have addressed the role of detection (Mehta et al. 2007). By increasing resources to detect invasive species, managers may increase their chances of finding a species at a smaller population level and making control potentially less expensive and more effective. However, detecting new invasive species is difficult and uncertain. Many factors reduce successful detection, such as low population densities which are prevalent in invasive species management. Detection of new invasive species at low population levels is often difficult unless they produce highly visible damages (Mehta et al. 2007).

CONCLUSION

The above presented research indicates a relationship between mussel life and feed habits, nutrient loading and cycling, ecological shifts, and the bioaccumulation of contaminants, all of which result in changes of water quality and a potential threat to the health of the ecosystem, including humans. Mussels are of unique ecological value as natural biological filters, food for fish and wildlife, and indicators of water quality. Also mussels as habitat generators provide habitat for a variety of aquatic life forms and habitat conditions suitable for other aquatic organisms. In summary, marine mussels have a high nutritive value as a result of their complex compositions, and freshwater mussels are a highly diverse, historically abundant and widespread group. They are ecologically important and, as a group, are at a critical stage for conservation because of their large number of imperiled species. Water quality is a key parameter in the health of an aquatic environment. Mussels have great scientific value as indicators of environmental health because they remain essentially in one place for a long time and require good water and sediment quality and physical habitat. Mussels can be exposed to contaminants through many routes, including overlying waters, sediments, pore water, fish hosts and food. The amount of certain pollutants in mussel tissue can be measured to determine the type and extent of water pollution in various aquatic environments. Mussel toxicity data can be useful for water quality management purposes. Consequently, preserving the water quality has become an international priority. Mussels transfer nutrients and energy from the water column to the sediments through their filtering activity and stimulate production across trophic levels. Mussels aggregate in water beds, controlling nutrient production and cycling, and invertebrate species composition and abundance. The mussels ability to alter the ecological community, shift chemical balances and thus water quality, highlights the need to determine how to control invasive mussels. Although the effects of invasive alien species on native species are well documented, the many ways in which such species impact ecosystems are still emerging. Invasive species are a leading cause of biodiversity loss and related economic damages. These amazing species are persistent subjects of science and biology. Scientists have argued that invasive mussels can be managed most cost effectively with greater investments in prevention.

ACKNOWLEDGMENTS

The authors acknowledge financial support from Serbian Ministry of Science grant III 43009.

REFERENCES

Ackerman, J. D.; Sim, B.; Ncholss, J. and Claudi, R. (1994). A review of the early life history of zebra mussels *(Dreissena polymorpha):* comparisons with marine bivalves. *Canadian Journal of Zoology*, 72, 1169–1179.

Anzano, J.; Lasheras, R. –J.; Bonilla, B.; Bonilla, A.; Lanaja, J.; Peribañez, M. A.; Gracia–Salinas, M. –J.; Anwar, J. and Shafique, U. (2011). Determination of trace metals by voltamperometry in zebra mussel (*Dreissena polymorpha*) employed as environmental bio–indicator. *Green Chemistry Letters and Reviews*, 4, 261–267.

Baker, S. M. and Levinton, S. J. (2003). Selective feeding by three native North American freshwater mussels implies food competition with zebra mussels. *Hydrobiologia*, 505, 97–105.

Barbiero, R. P.; Rockwell, D. C.; Warren, G. J. and Tuckman, M. L. (2006). Changes in spring phytoplankton communities and nutrient dynamics in the eastern basin of Lake Erie since the invasion of *Dreissena* spp. *Canadian Journal of Fisheries and Aquatic Sciences*, 63, 1549–1563.

Barnhart, C. and Neves, D. (2005). Overview of North American freshwater mussels. Mussels Toxicity Testing Workshop Hosted by U.S. EPA, Chicago, Illinois, 2005.

Bauer, G. (1988).Threats to the freshwater pearl mussel in Central Europe. *Biological Conservation,* 45, 239–253.

Bauer, A.; Østensvik, Ø.; Florvag, M.; Ørmen, Ø. and Rørvik, L. M. (2006). Occurrence of *Vibrio parahaemolyticus*, *V. cholerae*, and *V. vulnificus* in Norwegian Blue Mussels (*Mytilus edulis*). *Applied and Environmental Microbiology*, 72, 3058–3061.

Beliaeff, B.; O'Connor, T. P. and Claisse, D. (1998). Comparison of chemical concentrations in mussels and oysters from the United States and France. *Environmental Monitoring and Assessment*, 49, 87–95.

Bellotto, V. R. and Miekeley, N. (2007). Trace metals in mussel shells and corresponding soft tissue samples: a validation experiment for the use of *Perna perna* shells in pollution monitoring. *Analytical and Bioanalytical Chemistry*, 389, 769–776.

Bervoets, L.; Voets, J.; Covavi, A.; Chu, S.; Qadah, D.; Smolders, R.; Schepens, P. and Blust, R. 2005. Use of transplanted zebra mussels (*Dreissena polymorpha*) to assess the bioavailability of microcontaminants in Flemish surface waters. *Environmental Science Technology*, 39, 1492–1505.

Binelli, A.; Galassi, S. and Provini, A. (2001). Factors affecting the use of *Dreissena polymorpha* as a bioindicator: the PCB pollution in Lake Como (N. Italy). *Water, Air and Soil Pollution*, 125, 19–32.

Binelli, A. and Provini, A. (2003). DDT is still a problem in developed countries: the heavy pollution of Lake Maggiore. *Chemosphere*, 52, 717–723.

Boltovskoy, D.; Sylvester, F.; Otaegui, A.; Leites, V. and Cataldo, H.D. (2009). Environmental modulation of reproductive activity of the invasive mussel *Limnoperna fortunei*: implications for antifouling strategies. *Austral Ecology*, 34, 719–730.

Bringolf, B. R.; Cope, G. W.; Chris, B.; Eads, B. C., Peter, R.; Lazaro, R. P.; Barnhart, M. C. and Shea, D. (2007). Acute and chronic toxicity of technical–grade pesticides to glochidia and juveniles of freshwater mussels (Unionidae). *Environmental Toxicology and Chemistry*, 26, 2086–2093.

Brzozowska, R.; Sui, Z. and Kang, K. H. (2012). Testing the usability of sea mussel (*Mytilus* sp.) for the improvement of seawater quality–An experimental study. *Ecological Engineering*, 39, 133–137.

Carro, N.; Garcia, I.; Ignacio, M. and Mouteira, A. (2005). Influence of raft rope depth and biological parameters on polychlorinated biphenyl levels in cultured mussels at Galicia coast (north–western Spain), and distribution patterns of PCBS in raft mussel. *Fresenius Environmental Bulletin*, 14, 884–893.

Chou, C. L.; Paon, L. A.; Moffatt, J. D. and King, T. (2003). Selection of bioindicators for monitoring marine environmental quality in the Bay of Fundy, Atlantic Canada. *Marine Pollution Bulletin*, 46, 756–762.

Cohen, A.N. (2007). Potential distribution of zebra mussels (Dreissena polymorpha) and quagga mussels (Dreissena bugensis) in California, phase 1 report—A report for the California Department of Fish and Game, 14 p., accessed November 20, 2012, at http://www.sfei.org/bioinvasions/ Reports / 2007 Dreissena%20Potential% 20Distribution.pdf.

Croci, L.; Serratore, P.; Cozzi, L.; Stacchini, A.; Milandri, S.; Suffredini, E. and Toti, L. (2001). Detection of Vibrionaceae in mussels and in their seawater growing area. *Letters in Applied Microbiology*, 32, 57–61.

Culha, S. T.; Celik, M. Y.; Culha, M.; Karayucel, I. and Gundogdu, A. (2008). The trace elements in the raft cultivated mussels (*Mytilus galloprovincialis* Lamarck, 1819) from Sinop Peninsula, in the southern Black Sea. *Journal of Animal and Veterinary Advances*, 7, 1618–1623.

Darrigran, G. (2002). Potential impact of filter–feeding invaders on temperate inland freshwater environments. *Biological Invasions*, 4, 145–156.

Devier, M. H.; Augagneur, S.; Budzinski, H.; Le Menach, K.; Mora, P.; Narbonne, J. F. and Garrigeus, P. (2005). One–year monitoring survey of organic compounds (PAHs, PCBs, TBT), heavy metals and biomarkers in blue mussels from the Arcachon Bay, France. *Journal of Environmental Monitoring*, 7, 224–240.

Downing, J. A.; Van Meter, P. and Woolnough, D. A. (2010). Suspects and evidence: a review of the causes of extirpation and decline in freshwater mussels. *Animal Biodiversity and Conservation*, 33, 151–185.

Elliott, P.; Aldridge, D. C. and Moggridge, G. D. (2008). Zebra mussel filtration and its potential uses in industrial water treatment. *Water Research*, 42, 1664–1674.

England, G. C.; McGrath, T. P.; Gilmer, L.; Seebold, J. G.; Lev–On, M. and Hunt, T. (2001). Hazardous air pollutant emissions from gas–fired combustion sources: emissions and the effects of design and fuel type. *Chemosphere*, 42, 745–764.

Fang, Z.–Q.; Cheung, R. Y. H. and Wong, M. H. (2003). Heavy metals in oysters, mussels and clams collected from coastal sites along the Pearl River Delta, South China. *Journal of Environmental Sciences*, 15, 9–24.

Farrington, J. W. and Tripp, B. W. International Mussel Watch Project. Initial Implementation Phase. Final Report, NOAA Technical Memorandum, NOS ORCA 95, Silver Spring, MD, 1995.

Garvin, M. Cultural importance of freshwater mussels to Native American tribes: a field study of caged mussels. In Proceedings Summary Report, Mussel Toxicity Testing Procedures Workshop, Chicago (Illinois), USA, August 23–24, 2005; pp 21.

Goodyear, L. K. and McNeill, S. (1999). Bioaccumulation of heavy metals by aquatic macro–invertebrates of different feeding guilds: a review. *The Science of the Total Environment*, 229, 1–19.

Gosling, E. Genetics of Mytilus. In *The mussel Mytilus: ecology, physiology genetics and culture*; Gosling, E., Ed.; Elsevier: Amsterdam, 1992; pp 309–382.

Graczyk, K. T.; Lucy, E. F.; Tamang, L.; Minchin, D. and Miraflor, A. (2008). Assessment of human waterborne parasites in Irish river basin districts – use of zebra mussels (*Dreissena polymorpha*) as bioindicators. *Aquatic Invasions*, 3, 305–313.

Griebeler, E. M. and Seitz, A. (2007). Effects of increasing temperatures on population dynamics of the zebra mussel *Dreissena polymorpha*: implications from an individual–based model. *Oecologia*, 151, 530–543.

Gugliandolo, C.; Lentini, V.; Spano, A. and Maugeri, T. L. (2010). Conventional and molecular methods to detect bacterial pathogens in mussels. *Letters in Applied Microbiology*, 52, 15–21.

Gutierrez, J. L.; Jones, C. G.; Strayer, D. L. and Iribarne, O. O. (2003). Mollusks as ecosystem engineers: the role of shell production in aquatic habitats. *Oikos*, 101, 79–90.

Heinzow, B.; Mohr, S.; Ostendorp, G.; Kerst, M. and Körner, W. (2007). PCB and dioxin-like PCB in indoor air of public buildings contaminated with different PCB sources – deriving toxicity equivalent concentrations from standard PCB congeners. *Chemosphere*, 67, 1746–1753.

Helfrich, A. L.; Neves, J. R.; Weigmann, L. D.; Speenburgh, M. R.; Beaty, B.; Biggins, D. and Vinson, H. (2009). Help save America's pearly mussels. College of Agriculture and Life Sciences, Virginia Polytechnic Institute and State University. Publication 420–014 Sponsored by U.S. Fish and Wildlife Service Asheville Field Office, Asheville, North Carolina. http://pubs.ext.vt.edu/420/420-014/420-014.html. (accessed Dec 1, 2012).

Holt, E. A. and Miller, S. W. (2011). Bioindicators: Using organisms to measure environmental impacts. *Nature Education Knowledge*, 2, 1–5.

Jain, C. K. (2004). Metal fractionation study on bed sediments of River Yamuna, India. *Water Research*, 38, 569–578.

Joksimovic, D.; Tomic, I.; Stankovic, R. A.; Jovic, M. and Stankovic S. (2011). Trace metal concentrations in Mediterranean blue mussel and surface sediments and evaluation of the mussels quality and possible risks of high human consumption. *Food Chemistry*, 127, 632–637.

Jones, L. A. and Ricciardi, A. (2005). Influence of physicochemical factors on the distribution and biomass of invasive mussels (*Dreissena polymorpha* and *Dreissena bugensis*) in the St. Lawrence River. *Canadian Journal of Fisheries and Aquatic Sciences*, 62, 1953–1962.

Jovic, M.; Stankovic, A.; Slavkovic–Beskoski, L.; Tomic, I.; Degetto, S. and Stankovic, S. (2011). Mussels as a bio–indicator of the environmental quality of the coastal water of the Boka Kotorska Bay (Montenegro). *Journal of the Serbian Chemical Society*, 76, 933–946.

Jovic, M.; Onjia, A. and Stankovic, S. (2012). Toxic metal health risk by mussel consumption. *Environmental Chemistry Letters*, 10, 69–77.

Juhel, G.; Davenport, J.; O'Halloran, J.; Culloty, S. C.; O'Riordan, R. M.; James, K. J.; Furey, A. and Allis, O. (2006). Impacts of microcystins on the feeding behaviour and energy balance of zebra mussels, *Dreissena polymorpha*: A bioenergetics approach. *Aquatic Toxicology*, 79, 391–400.

Kamaruzzaman, B. Y.; Mohd Zahir, M. S.; Akbar John, B.; Jalal, K. C. A; Shahbudin, S.; Al–Barwani, S. M. and Goddard, J. S. (2011).Bioaccumulation of some metals by green mussel *Perna viridis* (Linnaeus 1758) from Pekan, Pahang, Malaysia. *International Journal of Biological Chemistry*, 5, 54–60.

Kulakovskii, E. E. (2000). Biological basis of mussel mariculture in the White Sea, *Ser. Issledovaniya fauny morei*, St. Petersburg. *Nauka*, 2000, 50, 58.

Kwon, T. D.; Fisher, S. W.; Kim, G. W.; Hwang, H. and Kim, J. E. (2006). Trophic transfer and biotransformation of polychlorinated biphenyls in zebra mussel, round goby, and smallmouth bass in Lake Erie, USA. *Environmental Toxicology and Chemistry*, 25, 1068–1078.

Lachowicz, L. S. Population biology of mussels (*Aulacomya maoriana*, *Mytilus galloprovincialis* and *Perna canaliculus*) from rocky intertidal shores in Wellington Harbour, New Zealand. Ph.D. dissertation, Victoria University of Wellington, 2005.

Lewtas, J. (2007). Air pollution combustion emissions: Characterization of causative agents and mechanisms associated with cancer, reproductive, and cardiovascular effects. *Mutation Resarch*, 636, 95–133.

Liu, J. H. and Kueh, C. S. (2005). Biomonitoring of heavy metals and trace organics using the intertidal mussel *Perna viridis* in Hong Kong coastal waters. *Marine Pollution Bulletin*, 51, 857–875.

Lockwood, L.B. and Somero, N.G. (2011). Transcriptomic responses to heat stress in invasive and native blue mussels (genus *Mytilus*). *Molecular Ecology*, 20, 517–529.

Lovell, J.S. and Stone, F.S. (2005). The economic impacts of aquatic invasive species: a review of the literature. National Center for Environmental Economics U.S. EPA, http://www.epa.gov/economics

Ludyanskiy, L. M.; McDonald, D. and MacNeill, D. (1993). Impact of the Zebra Mussel, a Bivalve Invader. *BioScience*, 43, 533–544.

MacIsaac, H. J. (1996). Potential and abiotic and biotic impacts of Zebra mussels on the inland waters of North America. *American Zoology*, 36, 287–299.

Mackie, L. G. and Schloesser D. W. (1996). Comparative biology of Zebra nussels in Europe and North America: An Overview. *American Zoologist*, 36, 244–258.

Markovic, J., Joksimovic, D. and Stankovic, S. (2012). Trace element concentrations in wild mussels from the coastal area of the southeastern Adriatic, Montenegro. *Archives of Biological Sciences, Belgrade*, 64, 265–275.

Magara, Y.; Matsui, Y.; Goto, Y. and Yuasa, A. (2001). Invasion of the non–indigenous nuisance mussel, *Limnoperna fortunei*, into water supply facilities in Japan. *Aqua – Journal of Water Supply: Research and Technology*, 50, 113–124.

MarLIN, Marine Life Information Network. (2006). Biodiversity and Conservation – Common mussel – Mytilus edulis. http://www.marlin.ac.uk/species/Mytilusedulis.htm (accessed Dec 1, 2012).

McMahon, R. F. and Bogan, E. A. (2001). Mollusca: Bivalvia. Ecology and Classification of North American Freshwater Vertebrates. Thorp, H. J. and Covich, A. P., Eds.; Academic Press: San Diego, CA, 331–429.

Mehta, V.S.; Haight, G.R.; Homans, R.F.; Polasky, S. and Venette, C.R. (2007). Optimal detection and control strategies for invasive species management. *Ecological Economics* 61 237–245.

Mills, E. L.; Rosenberg, G.; Spidle, A. P.; Ludyanskiy, M.; Pligin, Y. and May, B. (1996). A review of the biology and ecology of the quagga mussel (*Dreissena bugensis*), a second species of freshwater Dreissenid introduced to North America. *American Zoological Journal*, 36, 271–286.

Minguez, L.; Boiché, A.; Sroda, S.; Mastitsky, S.; Brulé, N.; Bouquerel, J. and Giambérini, L. (2012). Cross–effects of nickel contamination and parasitism on zebra mussel physiology. *Ecotoxicology*, 21, 538–547.

Minier, C.; Abarnou, A.; Jaouen–Madoulet, A.; Le Guellec, A. M.; Tutundjian, R.; Bocquene, G. and Leboulenger, F. (2006). A pollution–monitoring pilot study involving contaminant and biomarker measurements in the seine estuary, France, using zebra mussels (*Dreissena polymorpha*). *Environmental Toxicology and Chemistry*, 25, 112–119.

Moore, M. N.; Wade, T.; Wedderurn, R. J.; Lowe, D. M.; Balashov, G.; Buyukgun, H.; Daurova, Y.; Denga, Y.; Kostylev, E.; Mihnea, P. E.; Moncheva, S.; Tabagari, S.; Ciocan, C.; Ozkoc, H. and Depledge, M. H. The UNESCO/IOC "Black Sea Mussel Watch Pilot Study": Biological effects and contaminant residues. In: Black Sea Pollution Assessment; Mee, L. D., Topping, G., Eds.; Black Sea Environmental Series Vol 10, United Nations Development Programme, New York, 1998; pp 279–292.

Mosher, S. Biomarkers of Lead Exposure in the Freshwater Mussel *Elliptio complanata* for Assessing Transportation Related Impacts. Master of Science, Toxicology Raleigh, North Carolina, USA, 2008.

Mozley, A. Distribution and its Limits. An Introduction to Molluscan Ecology. Lewis and Co. Ltd.: London, England, 1–36, 1954.

Nalepa, F. T. An Overview of the spread, distribution, and ecological iImpacts of the Quagga mussel, *Dreissena rostriformis bugensis*, with possible implications to the Colorado River system. In Proceedings of the Colorado River Basin Science and Resource Management Symposium, Scottsdale (Arizona), USA, November 18–20, 2008; pp 113–122.

Namiesnik, J.; Moncheva, S.; Park, Y. S.; Ham, K. S.; Heo, B. K.; Tashma, Z.; Katrich, E. and Gorinstein, S. (2008). Concentration of bioactive compounds in mussels *Mytilus galloprovincialis* as an indicator of pollution. *Chemosphere*, 73, 938–944.

Normanno, G.; Parisi. A.; Addante, N.; Quaglia, N. C.; Dambrosio, A.; Montagna, C. and Chiocco, D. (2006). *Vibrio parahaemolyticus*, *Vibrio vulnificus* and microorganisms of fecal origin in mussels (*Mytilus galloprovincialis*) sold in the Puglia region (Italy). *International Journal of Food Microbiology*, 106, 219–222.

Norstrom, R. J. (2002). Understanding bioaccumulation of POPs in food webs. *Environmental Science and Pollution Research*, 9, 300–303.

NRCS, Natural Recourses Conservation Service. (2007). Native freshwater mussels. *Fish and Wildlife Habitat Management Leaflet*, 46, 1–16.

Oehlmann, J. and Schulte–Oehlmann, U. Molluscs as bioindicators. In: *Bioindicators and biomonitors – Principles, concepts and applications*; Markert, B. A., Breure, A. M., Zechmeister H. G., Eds.; Elsevier: Amsterdam, 2003; pp 577–635.

Oros, D. R. and Ross, J. R. M. (2005). Polycyclic aromatic hydrocarbons in bivalves from the San Francisco Estuary: spatial distributions, temporal trends, and sources (1993–2001). *Marine Environmental Research*, 60, 466–488.

Osuna–Lopez, J. I.; Frias–Espericueta, M. G.; Lopez–Lopez, G.; Zazueta–Padilla, H. M.; Izaguirre–Fierro, G.; Paez–Osuna, F.; Ruiz–Fernandez, A. C. and Voltolina, D. (2009). Cd, Pb and organochlorine pesticides of *Mytella strigata* (Pelecypoda: Mytilidae) of six coastal lagoons of NW Mexico. *Boletín de investigaciones marinas y costeras*, 38, 233–239.

Page, D. S.; Boehm, P. D.; Brown, J. S.; Neff, J. M.; Burns, W. A. and Bence, A. E. (2005). Mussels document loss of bioavailable polycyclic aromatic hydrocarbons and the return to baseline conditions for oiled shorelines in Prince William Sound, Alaska. *Marine Environmental Research*, 60, 422–436.

Pain, S.; Biagianti–Risbourg, S. and Parant, M. *Relevance of the Multixenobiotic Defence Mechanism (MXDM) for the biological monitoring of freshwaters – example of its use in zebra mussel*. In *Trends in water pollution research*; Livingston, J. V., Ed.; Nova Science Publishers: New York, NY, 2005; pp 203–220.

Pimentel, D.; Zuniga, R. and Morrison, D. (2005).Update on the environmental and economic costs associated with alien-invasive species in the United States. *Ecological Economics*, 52, 273– 288.

Potrykus, J.; Albalat, A.; Pempkowiak, J. and Porte, C. (2003). Content and pattern of organic pollutants (PAHs, PCBs and DDT) in blue mussels (*Mytilus trossulus*) from the Southern Baltic Sea. *Oceanology*, 45, 337–355.

Rahnama, R.; Javanshir, A. and Mashinchian, A. (2011). The effect of lead bioaccumulation on condition indices of Zebra mussel (*Dreissena polymorpha*) from Anzali wetland–Caspian Sea. *Turkish Journal of Fisheries and Aquatic Sciences*, 11, 561–568.

Rehder, H.A. *National Audubon Society Field Guide to North American Seashells.* Knopf: New York, NY, 845–853, 1981.

Ricciardi, A. (1998). Global range expansion of the Asian mussel *Limnoperna fortunei* (Mytilidae): Another fouling threat to freshwater systems. *Biofouling,* 13(2):97-106.

Rius, M.and McQuaid, D.C. (2006). Wave action and competitive interaction between the invasive mussel *Mytilus galloprovincialis* and the indigenous *Perna perna* in South Africa. *Marine Biology*, 150, 69–78.

Ruus, A.; Green, N. W.; Maage, A. and Skei, J. (2006). PCB–containing paint and plaster caused extreme PCB–concentrations in biota from the Sorfjord (Western Norway)–A case study. *Marine Pollution Bulletin*, 52, 100–103.

Sagoff, M. (1999). What's wrong with alien species? Report of the Institute for Philosophy and Public Policy, 19, 16–23.

Sasikumar, G.; Krishnakumar, P. K. and Bhat, G. S. (2006). Monitoring trace metal contaminants in green mussel, *Perna viridis* from the coastal waters of Karnataka, southwest coast of India. *Archives of Environmental Contamination and Toxicology*, 51, 206–214.

Schaffner, M.; Bader, H. P. and Scheidegger, R. (2009). Modeling the contribution of point sources and non–point sources to Thachin River water pollution. *Science of the Total Environment*, 407, 4902–4915.

Schernewski, G.; Stybel, N. and Neumann, T. (2012). Zebra mussel farming in the Szczecin (Oder) Lagoon: water–quality objectives and cost–effectiveness. *Ecology and Society*, 17, 1–13.

Selegean, J. P. W.; Kusserow, R.; Patel, R.; Heidtke, T. M. and Ram, J. L. (2001). Using zebra mussels to monitor *Escherichia coli* in environmental waters. *Journal of Environmental Quality*, 30, 171–179.

Sericano, J. L.; Wade, T. L.; Jackson, T. J.; Brooks, J. M.; Tripp, B. W.; Farrington, J. W.; Mee, L. D.; Readman, J. W.; Villeneuve, J.–P. and Goldberg, E. D. (1995). Trace organic contamination in the Americas: an overview of the US National Status and Trends and the International 'Mussel Watch' Programmes. *Marine Pollution Bulletin*, 31, 214–225.

Skarpheoinsdottir, H.; Ericson, G. and Svavarsson, N. K. (2007). DNA adducts and polycyclic aromatic hydrocarbon (PAH) tissue levels in blue mussels (*Mytilus* spp.) from Nordic coastal sites. *Marine Environmental Research*, 64, 479–491.

Skinner, A.; Young, M. and Hastie, L. *Ecology of the freshwater Pearl mussel*, Conserving Natura 2000 Rivers Ecology Series No. 2; English Nature, Peterborough, 2003.

Sousa, R.; Gutierrez, L. J. and Aldridge, C. D. (2009). Non–indigenous invasive bivalves as ecosystem engineers. *Biological Invasions*, 11, 2367–2385.

Stankovic, S.; Jovic, M.; Stankovic, A. R. and Katsikas, L. Heavy metals in seafood mussels. Risks for human health. In *Environmental chemistry for a sustainable world*; Lichtfouse, E., Schwarzbauer, J., Robert D., Eds.; Springer: Netherlands, 2011; pp 311–373.

Stankovic, S. and Jovic, M. (2012). Health risks of heavy metals in the mediterranean mussels as seafood. *Environmental Chemistry Letters*, 10, 119–130.

Stankovic, S. and Stankovic R. A. (2012). Bioindicators of toxic metals. The book series "Environmental Chemistry for a sustainable World", Volume 3, in press Springer.

Strayer, D. L.; Downing, A. J.; Haag, R. W.; King, L. T.; Layzer, B. J.; Newton, J. T. and Nichols, J.S. (2004).Changing perspectives on pearly mussels, North America's most imperiled animals. *BioScience*, 34, 429–439.

Sures, B. (2004). Environmental parasitology: relevancy of parasites in monitoring environmental pollution. *Trends in Parasitology*, 20, 170–177.

Tanabe, S.; Kan–Atireklap, S.; Prudente, M. S. and Subramanian, A. Mussel Watch: Marine pollution monitoring of butyltins and organochlorines in coastal waters of Thailand, Phillipiines, and India. In: Proceedings of the Fourth International Scientific Symposium "Role of Ocean Sciencs for Sustainable Development"; UNESCO/IOC/WESTPAC, 1998; pp 331–345.

Tanabe, S. Mussel Watch: Marine Pollution Monitoring in Asian Waters. Center for Marine Environmental Studies; Ehime University, Japan, 2000.

Tronczynski, J.; Munschy, C.; Moisan, K.; Guiot, N.; Truquet, I.; Olivier, N.; Men, S. and Furaut, A. (2004). Contamination of the bay of biscay by polycyclic aromatic hydrocarbons (PAHs) following the T/V "Erika" oil spill. *Aquatic Living Resources*, 17, 243–259.

Turner, K.; Wong, W. H.; Gerstenberger, S. L. and Miller, J. M. (2011). Interagency monitoring action plan (I–MAP) for quagga mussels in Lake Mead, Nevada–Arizona, USA. *Aquatic Invasions*, 6, 195–204.

USFWS (2007). U.S. Fish and Wildlife Service, Western Quagga Mussels, http://www.fws.gov/rubylake/pdf/quaggamussel.pdf.

Van Caneghem, J.; Block, C.; Van Brecht, A.; Wauters, G. and Vandecasteele, C. (2010). Mass balance for POPs in hazardous and municipal solid waste incinerators. *Chemosphere*, 78, 701–708.

Vanderploeg, H. A.; Liebig, J. R.; Carmichael, W. W.; Agy, M. A.; Johengen, T. H.; Fahnenstiel, G. L. and Nalepa, T. F. (2001). Zebra mussel (*Dreissena polymorpha*) selective filtration promoted Microcystis blooms in Saginaw Bay (Lake Huron) and Lake Erie. *Canadian Journal of Fisheries and Aquatic Science*, 58, 1208–1221.

Vaughn, C. C.; Spooner, E. D. and Galbraith, S. H. (2007). Context–dependent species identity effects within a functional group of filter–feeding bivalves. *Ecology*, 88, 1654–1662.

Vaughn, C. C.; Nichols, J. S. and Spooner, E. D. (2008). Community and food web ecology of freshwater mussels. *Journal of the North American Benthological Society*, 27, 409–423.

Webb, K.; Craft, C. and Elswick, E. (2008). The Evaluation of the freshwater western Pearl mussel, *Margaritifera falcata* (Gould, 1850), as a bioindicator through the analysis of metal partitioning and bioaccumulation. *Northwest Science*, 82, 163–173.

WHO, World Health Organization. Persistent organic pollutants: Impact on child health. WHO Document Production Services, Geneva, Switzerland, 2010.

Wilson, A. (2003). Effects of zebra mussels on phytoplankton and ciliates: a field mescosm experiment. *Journal of Plankton Research*, 25, 905–915.

Wua, R. S. S.; Chan, A. K.; Richardson, B. J.; Au, D. W. T; Fang, J. K. H; Lam P. K. S. and Giesy, J. P. (2008). Measuring and monitoring persistent organic pollutants in the context of risk assessment. *Marine Pollution Bulletin*, 57, 236–244.

Wu, Y.; Bartell, M.S.; Orr J.; Ragland, J. and Anderson, D.(2010). A risk-based decision model and risk assessment of invasive mussels. *Ecological Complexity* 7, 243–255.

Zanatta, D. T. and Murphy, R. W. (2006). Evolution of active host–attraction strategies in the freshwater mussel tribe Lampsilini (Bivalvia: Unionidae). *Moleuclar Phylogenetics and Evolution*, 41, 195–208.

Zhou, Q.; Zhang, J.; Fu, J.; Shi, J. and Jiang, G. (2008). Biomonitoring: An appealing tool for assessment of metal pollution in the aquatic ecosystem. *Analytica Chimica Acta*, 606, 135–150.

Zotin, A. A. and Ozernyuk, D. N. (2004). Growth characteristics of the common mussel *Mytilus edulis* from the White Sea. *Biology Bulletin*, 31, 377–381.

Zuiganov, V.; Zotin, A.; Nezlin, L. and Tretiakov, V. The freshwater pearl mussels and their relationships with salmonid fish; VNIRO Publishing house: Moscow, 1994.

In: Mussels: Ecology, Life Habits and Control
Editors: Jarek Nowak and Michal Kozlowski
ISBN: 978-1-62618-083-3

Chapter 2

LIFE AND INTERACTION OF HYDROBIOLOGICAL ORGANISMS (*MYTILUS SP.*) WITH HARMFUL ALGAL BLOOMS: COMPOSITION, DISTRIBUTION AND METABOLISM OF HYDROPHILIC AND HYDROPHOBIC TOXINS IN THE AUSTRAL PACIFIC REGION

Carlos García[1,2,*] and Néstor lagos[1]
[1]Membrane Biochemistry Laboratory, Physiology and Biophysics Program, Faculty of Medicine, Universidad de Chile, Santiago, Chile
[2]Environment Department, Foundation for Human and Environmental Sciences Research. Santiago, Chile

ABSTRACT

For years, marine resources have been an important part of human development, given the high nutritional contribution they own. Due to requirements and needs exceeded natural inputs globally, man developed aquaculture, in order to regulate the imbalance between nutritional requirements and natural inputs, in which *Mussels sp*. are the most used species.

Under natural conditions, there are several and wide varieties of species of mussels, with multiple sea habitats: sandy, rocky and stratified bottoms in the water column in rocky strata, which allows them to assimilate a varied diet of micronutrients mostly constituted by phytoplankton. Although, depending on their habitat, mussels can filter wastes from other species such as faeces, pseudofaeces, microcellular debris or compounds from anthropogenic inputs.

From the many species of phytoplankton, about 40 of them are known by producing phycotoxins through blooms, being *Alexandrium catenella* and *Dinophysis sp.* the dinoflagellates'species associated with the production of toxins in the South Pacific Ocean. These dinoflagellates are filtered and accumulated by hydrobiological organisms,

* Corresponding author Email: Carlos Garcia, cgarcia@med.uchile.cl.

such as mussels, that accumulate toxins in the digestive glands, where are subjected to numerous reactions of biotransformation mediated by enzymatic, nonenzymatic pathways or by bacteria present in bivalves, which extends the toxic spectrum of the assimilated toxins, an event that is magnified by the interaction of vector organisms that expand the toxic forms through the trophic chain.

Paralytic shellfish toxins (STX isomers) and lipophilic toxins (DSP-toxins, AZAs, PTXs and YTXs), constitute the group of toxins that affect mussels from natural origin and aquaculture the most. These kind of toxins are accumulated in the tissues of mussels where are retained according to their chemical affinity and modified into more toxic forms (C2→GTX3) and/or thermodynamically more stable forms (GTX3→GTX2; 7-epi-PTX-2-SA→PTX-2-SA), allowing a harmlessness to accumulated toxins by mussels in some cases (DTX1→DTX3; AZA3→AZA5).

Since blooms associated with PSP-toxins and lipophilic toxins may occur simultaneously, different species of mussels and those species related to their habitat can accumulate and/or transform the wide spectrum of toxins, generating a toxin profile and variable toxicity in their tissues, which last according to the exposure time to the bloom and the purification pathway of toxins from mussels.

In this paper, the main events associated with paralytic shellfish toxins and lipophilic toxins as well as their interaction in the habitat of *Mussels sp.* in the South Pacific Ocean are presented.

1. Introduction

Mytilids, internationally known as "mussels", are bivalves, filter-feeding molluscs that belong to the family Mytilidae, bivalvia class, which pump seawater through their gills to filter out food items and other suspended particles. They release filtered water back into the environment. Materials retained by a mucous material on the gills—including tiny organisms, detritus, suspended sediment, and contaminants—are eaten and then assimilated into their tissues or passed through their bodies undigested. By filtering, they remove the phytoplankton and inorganic particles from the water column, thereby reducing turbidity, so turbidity reduction is directly proportional to the abundance of bivalves [1].

In the southern Pacific coast, mytilids are distributed from Perú to the Strait of Magellan and, in the Atlantic Coast, along the coast of Argentina to Brazil.

From the 70 species of mytilids described for the South Pacific Region, three species are economically important resources in Chile: Blue mussels (Chorito - *Mytilus Chilensis*), Giant mussels (Choro zapato - *Choromytilus Chorus*) and Ribbed mussels (Cholga - *Aulacomya ater*), for which there are closures during a time of the year to allow controlling the extraction from natural beds and thus ensuring their availability in a future time [2].

Under natural conditions, there are diverse and wide varieties of mussels' species with multiple habitats at sea, such as sandy, rocky and stratified bottoms in the water column in rock strata together with species such as *Pyurapreaputiali*s, *Lessonia nigrescens* and *Perumytilus purpuratus*, which allow them to absorb a varied diet of micronutrients consisting mainly of phytoplankton, and depending on their habitat, wastes from other mussels' species such as pseudofaeces, micro-cellular debris or compounds from anthropogenic inputs, too. These baseline nutritional intakes are added to the ascending variable produced by climate change that could be a predominant factor in the development of marine resources from natural beds or aquaculture development products.

For years, natural marine resources have been an important part in human development given the high nutritional content they have. Their contribution has been so significant that requirements and needs have exceeded natural contribution worldwide, going from sustainable resources to endangered natural resources or, directly, resources under threat of extinction. In order to adjust the imbalance between nutritional requirements and natural inputs, man developed aquaculture; through the latter, it is possible to appease natural damage caused to natural resources and to allow an adequate separation between natural resources as such from those resources coming from farming centers, thus preventing the extinction of more mussels species of great commercial interest. About 85% of the global production comes from aquaculture activities, with the exception of a few countries where the extraction is strictly controlled, natural beds have been overexploited. The supply of this kind of global resources is based on the stability of the activity and the ability of farmers to increase their crops and to improve products [3].

The phytoplankton biomass increase associated with the nutrient-enriched input is beneficial for filter feeders bivalves. Nevertheless, N and P inputs from anthropogenic sources alter the nutrients ratio (16:1) to very high levels (70:1) by nutrients from agriculture through rainwater. In nutrient enriched systems with high levels of primary production, intra- and interspecific competition for food is minimal when the abundance of bivalves is at low or intermediate levels. However, in less productive systems with limited water circulation and/or high levels of bivalves' biomass, intra- and interspecific competition may occur between stocks from natural bivalves and those produced by aquaculture [4].

Additionally, organic imbalance (N:P) generated in some areas globally, produces a variability in marine phytoplankton communities, generating in many cases an exponential growth of these species, which, even though are still a nutritional content for multiple species of mussels, generate high commercial and human impact, by being associated with the production of toxins capable of killing people [5].

To date, 5,000 phytoplankton species have been described, from which, under certain circumstances, 300 species have a high proliferation rate resulting in a high density of microalgae in the sea, a condition called "bloom". The circumstances or factors triggering these "blooms" are still not completely determined, but what is certain is that specific changes of climatic and hydrobiological conditions play an important role in the conditions leading to a "bloom". These "blooms", in general, generate beneficial conditions for the development of aquaculture and marine biology. However, about 40 species, out of the 300 already mentioned found in phytoplankton, are dinoflagellates and diatoms that -under certain circumstances- produce phycotoxins (marine toxins). The abundance of this kind of toxic phytoplankton range from a few hundreds of thousands up to several million per liter in the sea, reaching maximum levels called "Harmful Algal Blooms" (HABs) [6,7]. HABs are produced by a group of cells (dinoflagellates, cyanobacteria and diatomae) that, due to different natural environmental or anthropogenic conditions, whose population may prevail over other microplankton species and produce strong toxins, can dramatically affect marine fauna, mammals and people. Authorities of inhabited zones where HABs are detected, react closing large sea areas to fishing and halting harvesting activities in zones where farmed and wild molluscs and bivalve are found (Figure 1) [8,9]. Phycotoxins can be accumulated in several marine species, such as fishes, crabs an every kind of filtering bivalves. Mussels, because of their high rate of filtering, can accumulate great amounts of phycotoxins in their digestive glands, with no adverse effects. However, when highly contaminated mussels are eaten by animals or persons, can produce severe intoxications. The worldwide number of phycotoxins induced intoxications per year is about 60,000 cases. Additionally, there is a strong impact on ecosystems, environment and economy, as sequels of the negative effect on tourism, recreation and miticulture and aquaculture industries [10].

Mussels coming from aquaculture facilities can alter the nutrients in the water column, which occurs with visibility in all seasons except summer, where phytoplankton grow faster, not being able to control the phytoplankton biomass. Moreover, water flows increase the seston flow through the crop sites, thus ensuring an adequate supply of food to bivalves [11].

In Europe, annual estimated losses in the tourism area, as consequence of HABs, are approximately €$ 700 million and 116 million on miticulture [12]. In order to prevent intoxications caused by eating phycotoxins contaminated shellfish and to mitigate their harmful effects, international agencies have produced regulations, legislation and monitoring programs [13]. Health risk prevention, as well as industry need for monitoring phycotoxin presence within several toxin carrying vectors, make necessary to develop a potentially toxic microalgae watch program. Several phycotoxin detection methods have been developed, many of which allow detection of phycotoxins within shellfish far below dangerous levels. Those methods challenge many difficult issues, such as insertion of toxin within complex carrying matrix and the presence of mixtures of structurally related analogs [14].

The basic contribution of toxins provided from harmful phytoplankton is varied, depending on the involved species and the areas where the bloom is produced at a global level, considering besides that different species of toxin-producing microalgae may produce simultaneous blooms, thus, causing that a same species of hydrobiological organisms, from both natural beds or aquaculture facilities, may accumulate more than one variety of toxins resulting in a toxic profile and variety of toxicity which is increased as they get involved in the trophic chain. Nevertheless, mussels do not only accumulate these toxins inside their bodies, but also in their metabolic pathways and pathways mediated by enzymes; they make key modifications in order to decrease their toxicity towards them, which often cannot be extrapolated to humans, for whom these chemical modifications may result more toxic.

This impact of natural origin do not only affects natural resources, but also resources from the development of aquaculture by adding, then, economic and social development effects in local communities that live from this activity [15]. By affecting the development of aquaculture, HABs have added another variable to the toxic composition that mussels have, since to the contribution of basal toxins and to those derived as a result from biotransformation processes caused by mussels, those produced by the different stages of productive developments (canned or frozen products) are also added, which when subjected to multiple physical-chemical variables, further expanding the toxins forms that mussels have, they become into more toxic forms or from variants that molluscs have transformed into innocuous forms to toxic forms again [16].

Based on the chemical properties of the toxins found in phytoplankton and shellfish, those have been classified in two classes: hydrophilic and lipophilic toxins. Toxins associated to Paralytic Shellfish Poison (PSP) are hydrophilic and share a common trait: they are of low molecular weight (about 500 Dalton) and toxins associated to lipophilic toxins, Diarrhetic Shellfish Poisoning (DSP), Azaspiracid Shellfish Poisoning (AZP), Pectenotoxins (PTX) and Yessotoxins (YTX) share two common traits: 1) Their molecular weight is between 600 and more than 2000 Dalton; and 2) they have a lipophilic chemical nature [10].

Figure 1. Summary of environmental impacts of Harmful Algal Blooms (HABs) on carnivorous, shellfish (mussels) and miticulture from the South Pacific Ocean.

2. Paralytic Shellfish Poison-Toxins

Within the toxins produced by HABs, paralytic shellfish poisons (PSP), a group of neurotoxins with high affinity with voltage-dependent sodium channels that cause muscle paralysis by blocking the nervous impulse, can be found [17,18]. These toxins are constituted by a unit called imidazoline that according to modification of some of its functional groups can be divided in three groups: Carbamoyl-toxins, N-sulfocarbamoyl-toxins and decarbamoyl-toxins. Up to now, over 57 isomers related to this group of PSP-toxins (PSTs) have been identified, each of them possessing different toxic capabilities (>carbamoyltoxins >decarbamoyltoxins >N-sulfocarbamoyltoxins (C-toxins)), but all of them have the same blocking action on the sodium channels in the nerve cells, preventing conductance of the synaptic signal along neuron leaving as net result both paralysis and death from respiratory failure possibly [19,20].

Among the most important PSTs congeners, gonyautoxins (GTX), (GTX4/GTX1 and GTX3/GTX2), neosaxitoxin (neoSTX) and saxitoxin (STX) corresponding to carbamoyltoxins groups; C1, C2, C3, C4, GTX5 and GTX6 corresponding to N-sulfocarbamoyltoxins groups and dcSTX, dcneoSTX, dcGTX3/dcGTX2 and dcGTX4/dcGTX1 corresponding to decarbamoyltoxins-group are highlighted (Figure 2) [21,22].

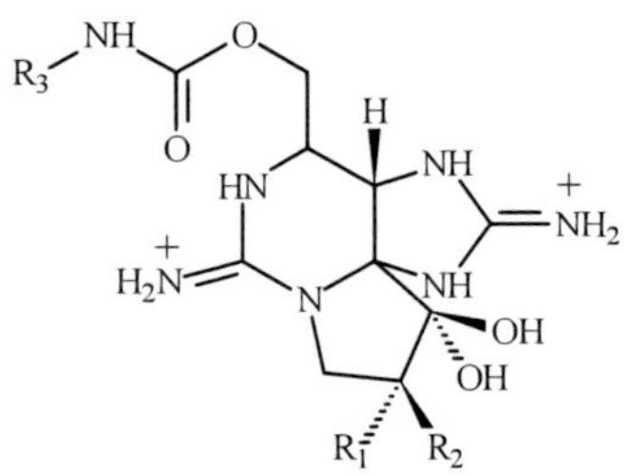

N1-HType

	R_1	R_2	R_3
STX	H	H	H
GTX5	H	H	SO_3^-
GTX2	OSO_3^-	H	H
GTX3	H	OSO_3^-	H
C1	OSO_3^-	H	SO_3^-
C2	H	OSO_3^-	SO_3^-

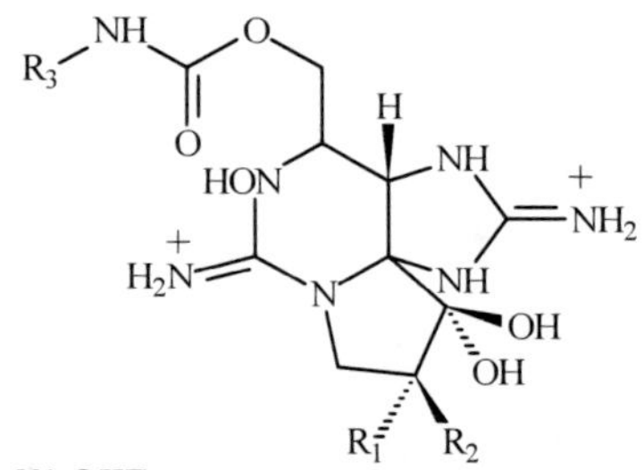

N1-OHType

	R_1	R_2	R_3
neoSTX	H	H	H
GTX6	H	H	SO_3^-
GTX1	OSO_3^-	H	H
GTX4	H	OSO_3^-	H
C3	OSO_3^-	H	SO_3^-
C4	H	OSO_3^-	SO_3^-

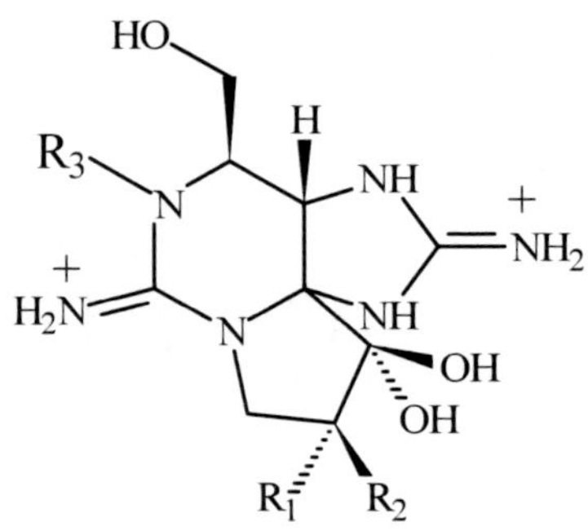

Decarbamoyl-PSP-Toxins

	R_1	R_2	R_3
dcSTX	H	H	H
dcneoSTX	H	H	OH
dcGTX1	OSO_3^-	H	OH
dcGTX2	OSO_3^-	H	H
dcGTX3	H	OSO_3^-	H
dcGTX4	H	OSO_3^-	OH

Figure 2. Chemical structures of Paralytic Shellfish Toxins. Saxitoxin = STX; neosaxitoxin = neoSTX; Gonyautoxins 1,2,3,4,5,6 = GTX1, GTX 2, GTX 3, GTX 4, GTX 5, GTX 6 ; decarbamoyl toxins = dc.

The alternative method as a replacement for the current AOAC biological method (Mouse bioassay) for determining paralytic shellfish toxins (PSTs) in mussels *(Mytilus edulis),* softshell clams *(Mya arenaria),* sea scallops (*Placopectin magellanicus*), and oysters *(Crassostrea virginica)* corresponds to the HPLC with post-column oxidation (PCOX) method, which allows determining the concentrations of individual PSTs, including saxitoxin, neosaxitoxin, gonyautoxins-1 to -5, decarbamoyl-gonyautoxins-2 and -3, decarbamoyl-saxitoxin, and *N*-sulfocarbamoyl-gonyautoxin-2 and -3, in edible tissues, as well as the total toxicity of the test samples above 0.10 mg STX dihydrochloride equivalents per kilogram of tissue (mg STX·diHCl/kg) [23]

By filtering water, mussels accumulate PSP-producing dinoflagellates, thus acquiring high toxicities of toxins in their digestive glands and that could cause death when consumed by humans. Current seafood regulations set the maximum acceptable level of PSTs in shellfish at 80 μg STX eq 100 g^{-1} of tissue in most countries [20].

Symptoms associated with PSP-toxins intoxications include tingling sensation of the lips, mouth and tongue, numbness of extremities, paresthesias, weakness, ataxia, floating/dissociative feeling, nausea, shortness of breath, dizziness, vomiting, headache, dysphagia and dysarthria, which start within 30 minutes after eating contaminated mussels (≈120 μg STX equivalent per person), time which basically depends on the concentration of mussels, that could lead to death of a person between 3 and 4 hours once the symptoms appear (≈ 400 a 10,000 μg STX equiv per person) [24-26].

PSP-toxins are produced by marine dinoflagellates of the *Alexandrium, Pyrodinium* and *Gymnodinium* genres (*Alexandrium tamerense, Alexandrium fundyense, Alexandrium catenella, Gimnodinium catenatum* and *Pyrodinum bahamense)* [6], also by certain species of cyanobacteria such as *Anabaena circinalis, Anabaena limmermernii, Aphanizomenon gracile, Aphanizomenon issatschenkoi, Cylondrospermopsis raciborskii, Lynybya wollei, Planktotrix sp.* and *Rivularia sp.* [27,28]. Since PSP-toxins are secondary metabolites with no apparent essential functional role for dinoflagellates, at limited concentrations of nitrogen (N), both the nitrogen pool and the intracellular energy are used for basic cell function processes, therefore, limited nitrogen and energy requirements do not favor the synthesis of toxins [29].

Alexandrium is an important dinoflagellates genus widely distributed worldwide. In South America, seasonal blooms of several genera of *Alexandrium* have been registered (*Alexandrium tamerense, Alexandrium catenella, Alexandrium minutum* and *Alexandrium tamayavanichii*), where the total toxicity of a dinoflagellates cell depends on the absolute toxin concentration and the relative toxin composition [30-32].

For example, blooms associated with *Alexandrium sp.* are typically very dynamic and from very different toxicities in each of the identified regions, which is directly associated with the specific response to environmental conditions to which it is exposed in the different regions [27,33]. Through this, it has been shown that light, temperature, nutrients and salinity impact on the photosynthesis of *Alexandrium sp.* and hence, on its growth and toxicity [33].

Preliminarily, it has been determined that dinoflagellates of *Alexandrium sp.* tend to show abnormal shapes (small shape and size) at temperatures equal or above 25 ° C, which results in the cyst formation, thus declining cell growth. These cysts contain little or no Chlorophyll, and the nuclear material becomes more compact than in vegetative cells, determining that the optimum growth conditions for this species average 20 ° C and a salinity of 30 PSU, allowing them to reach an average toxins content of 244 fmol per cell [33].

For its part, the amount of toxins associated with *Alexandrium sp.* is variable and dependent on the geographical area in which it has been identified, so, for example, species obtained in Argentina, Uruguay and Brazil average a toxicity of 2.3 – 65.6 pg STX equiv. cell^{-1}, where the differences between the toxic profile of dinoflagellates identified in the South Pacific and those in the South Atlantic Ocean, correspond to the presence of GTX5 [34,35].

Regarding the composition of the toxic profile in the different species of *Alexandrium sp.*, variability is directly related to the season in which they are assessed. Thus, toxic profiles (% mol) obtained in spring predominate β-epimers (C2, GTX4, GTX3) and in autumn α-epimers (C1, GTX1, GTX2), this is since during periods unfavorable to blooms, toxins that form part of the profile tend to transform (epimerization) to more stable forms [29]. The before mentioned is directly related to the data obtained from cyts, in which the profiles are predominated by forms 11-α-hydroxysulfate epimer, clearly indicating that the formation of these epimers in cysts are correlated with nitrogen-limited states [36].

The determination of the toxins content (fmol cell^{-1}) and toxicity (pg STX equiv cell^{-1}) in cysts, dinoflagellates and mussels require special attention of pH range and temperature to which they are subjected, since unstable conditions generate the transformation of N-sulfocarbamoyl toxins into their corresponding carbamoyl-derivatives (C1 $\rightarrow$ GTX2; C2 $\rightarrow$ GTX3; C3 $\rightarrow$ GTX1; C4 $\rightarrow$ GTX4) thus, affecting and generating a misinterpretation of the associated toxins profile [33,37].

Blooms associated with *Alexandrium catenella* in the South Pacific Ocean mainly occur between January and March, where the influence of the oscillation in the ocean plays an important role since it modifies the characteristics of the general circulation of inland waters and temperature of the water column, promoting or inhibiting the blooms of *Alexandrium catenella*. Additionally, the increase in the spatial distribution of the species' blooms appears to be strongly influenced by the drift of surface waters, caused by winds and by characteristics of inland waters circulation [38]. Therefore, blooms associated with this species in the southern region, are closely related to optimal conditions of population growth and / or germination of cysts, such as temperature, salinity, nutrients, thermal stratification, influence of winds and currents [39,40].

Blooms and their abundance of *Alexandrium sp.* suggest a dynamic highly dependent on the existence of cysts bank and on its germination process, since it is not possible to explain a bloom only through the vegetative process, due to the low specific growth rates associated with *Alexandrium catenella* (0,47 to 0,55 d^{-1}) [40]. So, the presence of cysts is more important than the hydrobiological and meteorological variations for the success of a bloom [41].

Most of the blooms associated with *Alexandrium sp.* have annual records in which the population dynamics of the species is related to the presence of cysts banks that are responsible for the increase in population [42]. However, between 1996 and 2002 in inland waters of northwestern Patagonia (Chile), a bloom of *Alexandrium catenella* with a biannual pattern was registered, establishing that cysts of this species would need a period of dormancy before germination, which would explain the biannual pattern observed in northwestern Patagonia [38,43].

In the South Pacific Ocean, profiles of *Alexandrium catenella* are characterized by being composed in about 70% by N-sulfocarbamoyl derivatives (C1/C2, GTX5), 24 % carbamoyltoxins (GTX4/GTX1) and in smaller proportion (≈ 3%) GTX3/GTX2, GTX6, neoSTX and STX; in trace levels dcGTX3/dcGTX2 and dcSTX may be evidenced [35].

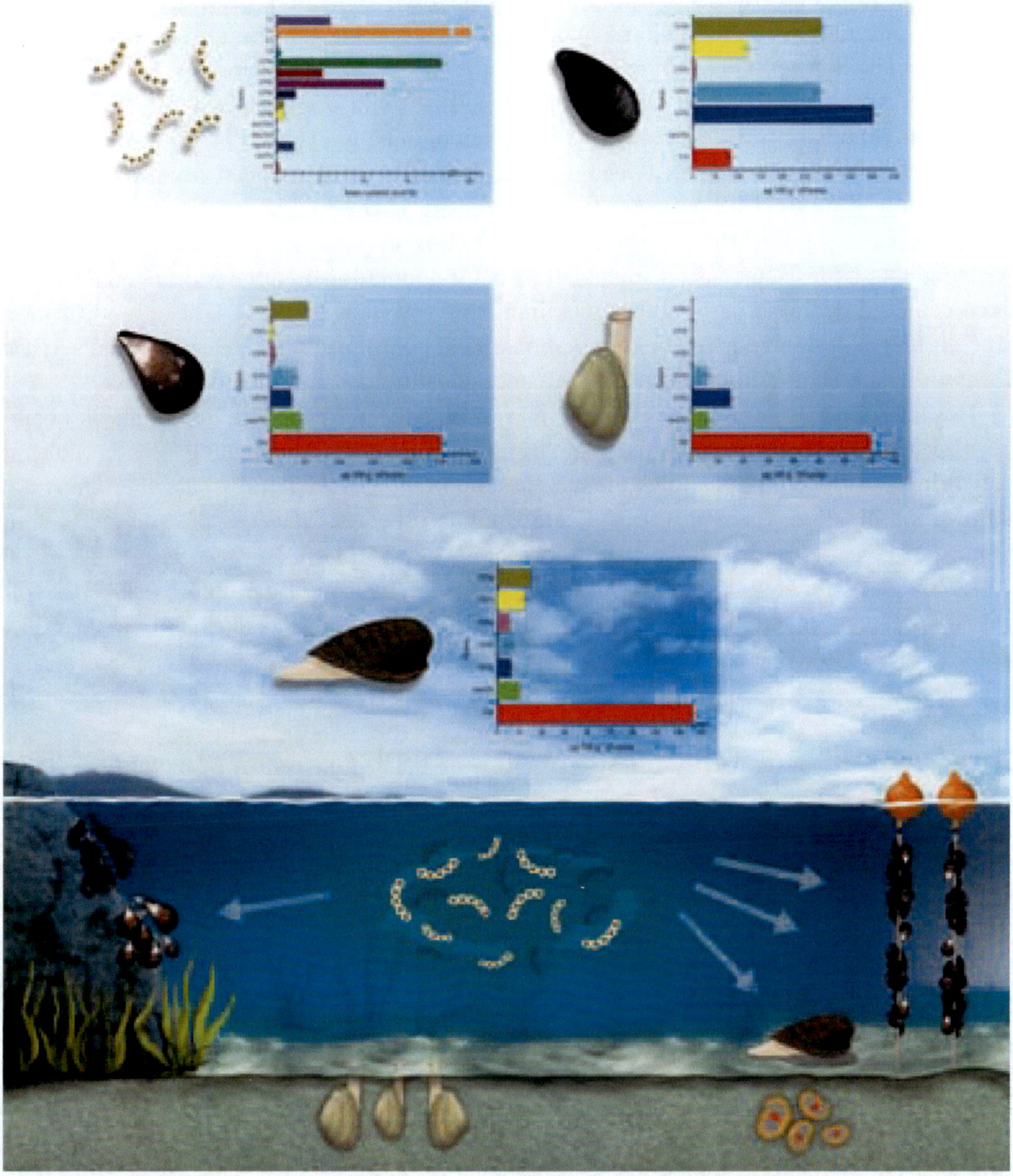

Figure 3. Comparison of PSP-toxin profiles of the dinoflagellate *Alexandrium catenella*, bivalve species Blue mussels (*Mytilus chilensis*) Ribbed mussels (*Aulacomya ater*), Razor Clam (*Tagelus dombeii*) and carnivorous gastropods Loco (*Concholepas concholepas*) from the South Pacific Ocean.

For its part, bivalves filter the PSP-toxin producing dinoflagellates, acting as carriers after the accumulation of this kind of toxin in their digestive tissues and glands, transmitting these toxins through the trophic chain to higher species and carnivorous gastropods such as sea mammals and fish, increasing the variability of toxic profiles and toxins associated to human poisonings [27, 44].

Therefore, comparative studies of profiles identified in dinoflagellates, mussels and poisoned people have traced the metabolic pathways of these PSP-toxin isomers follow -

under the action of different enzymatic (desulphation, reduction or hydrolysis) or non-enzymatic reactions (epimerization)- when they have been transmitted between various mussel species (Figure 3) [45, 46].

Furthermore, it has been established that shellfish with constant exposure to HAB are more sensitive to harmful algae compared to those continually exposed [47], so shellfish non-exposed to blooms show reduced feeding rates on toxic cells, reduced toxin accumulation rates and lower survival [48,49]. For example, Clam *Mya arenaria* corresponds to a more resistant species to PSP-toxins, accumulating toxins at high speed, where studies have concluded that its resistance is attributable to a natural mutation on a the sodium channel gene which causes a decrease of ≈ 1,000-times in affinity by STX [17, 50]. These modifications may correspond to spontaneous mutations and chromosomal rearrangements (base substitution, single base frame-shift and sequence substitutions mutations) [51].

This sensitivity of some species allow mussels to develop resistance to blooms associated with PSP-toxins, which added to its high filtration capacity (filter-feeding mussels) and its relative insensitivity to PST allows them to be important PST vectors, where the removal of toxins from bivalves can vary considerably among species, showing in some cases the presence of PST even until two years after the intake of PST [31,52].

At first, it was interpreted that mussels and vectors were not affected by HAB beyond their toxic effects in the trophic chain, but it has been established that these toxic effects clearly induce on the behavior, metabolism and survival of the involved species [53].

2. A. Composition and Biotransformation of PSP-Toxins on Mussels

By defining the toxic composition of mussels, it is concluded they usually have low levels of N-sulfocarbamoyl-toxins (C-toxins), but high levels of carbamates (GTX), where the difference of toxin profiles among species is associated to the transformation speed of toxins and the different retention capabilities of them.

Furthermore, enzymatic conversion of PST into toxic forms such as decarbamoyl derivatives is not commonly found in mussels [54]. Additionally, it has been established that some bacteria (*Pseudomonas sp.*) characteristic in different species of mussels and shellfish may transform via reduction some toxins such as GTX4/GTX1 into GTX3/GTX2 [55].

In unified systems where a variable habitat of aquatic organisms exists, differences in toxic profiles for rocky strata and/or hard bottoms mytilids (*Mytilus chilensis*, *Aulacomya ater* and *Choromytilus chorus*) which are characterized by having a variability of toxic profiles that make the distinction among species, can be clearly defined. All of them are characterized by having GTX4/GTX1:GTX3/GTX2 isomers in a ratio of 1:0.8, although despite this relationship, species such as Ribbed mussels (*Aulacomya ater*) and Giant mussels (*Choromytilus chorus)* are characterized by presenting profiles dominated by STX from a toxic point of view, demonstrating the high capacity of biotransformation that toxins can be exposed to through the high rates of enzymatic conversion they have (glutatione or sulfotranferasas) and/or to high bacteria levels they may have that favour these pathways of chemical transformations, characteristic of reductive eliminations in N_1-hydroxyl and C_{11}-hydroxysulfate groups.

Meanwhile, even when the habitat conditions are the same of the bivalves' under studies, *Mytilus chilensis* presents profiles with a same relationship among the identified GTXs, which clearly dominate the toxicity of the species (92%), showing no presence of neoSTX and low levels of STX (7%), so that its profile is more associative to the species responsible for HAB (*Alexandrium catenella*) in the South Pacific zone. Additionally, upward transformation of GTX4 into thermodynamically more stable chemical forms such as GTX1, characteristic in profiles of mussels, becomes evident.

Furthermore, in mussels, high levels of GTX3/GTX2 (ratio 0.8: 1) with predominance of the more stable enantiomer, GTX2, which occurs due to the high degree of biotransformation of C1/C2 toxins firstly by epimerization and secondly by hydrolisis into more stable forms are evident, a condition that is accentuated in periods of duration of mussels where the GTX3/GTX2 epimers predominate the toxic profiles (Figure 3) [35, 46].

Among the transformations in mussels, the conversion of toxins from less stable isomers (C2, GTX4, GTX3) β-epimers into chemically more stable forms (C1, GTX1, GTX2) α-epimers and from gonyautoxins (GTX) into neoSTX and /or STX is evident [45,56]. So, these enzymatic and nonenzymatic chemical transformations are directly related to compartmentalization of toxins in mussel's tissues.

Thus, in the ratio of toxins provided by dinoflagellates (*Alexandrium catenella)* (C2/C1 and GTX1/GTX4) with predominance of β-epimers, when filtered by mussels, they are rapidly transformed via epimerization into thermodynamically more stable forms, becoming α-epimers dominant in the profile.

Then, in subsequent stages, initially non-modified C2/C1 forms, and which may be identified in low concentrations in mussles, when stored in their digestive glands, are transformed into other toxic forms such as GTX4/GTX1 and these, in turn, are transformed into GTX3/GTX2 in later stages of a bloom or in purification stages of mussels, corresponding to more stable toxic forms and, therefore, they tend to stay longer in mussels. Besides, in this stage, toxic forms more associated to enzymatic pathways such as the identification of isomers of neoSTX and STX which toxic forms are more stable and of slow depuration, are evidenced.

It should be noted that each stage prior to the chemical modification has the ability to be purified or removed by mussels at trace level or larger amounts, which largely depends on the oceanographic conditions or availability of food (Figure 4A).

The main accumulation's tissues of PSP-toxins in mussels correspond to digestive glands (hepatopancreas), characterized by having a ratio of 10:1 with respect to non visceral tissues such as gonads, gills, mantle and adductor muscle (Figure 4B).

This distribution is variable in time, tending to change the profile of predominance of toxins in each tissue, given by the biotransformation that toxins undergo and by different removal rates of toxins from the different compartments.

Other species of bivalves which share the habitat of mussels, (*Venus antiqua*, *Gari solida* and *Tagelus dombeii*), tend to differ from the previously described species, since they are from soft bottom habitat (sand), where these species are constantly buried or semi-buried (infaunal), so their diet (filter-feeders) is based on wastes of the water column, whether microlagae (toxic dinoflagellates), faeces and/or pseudofaeces from other filter-feeding organisms.

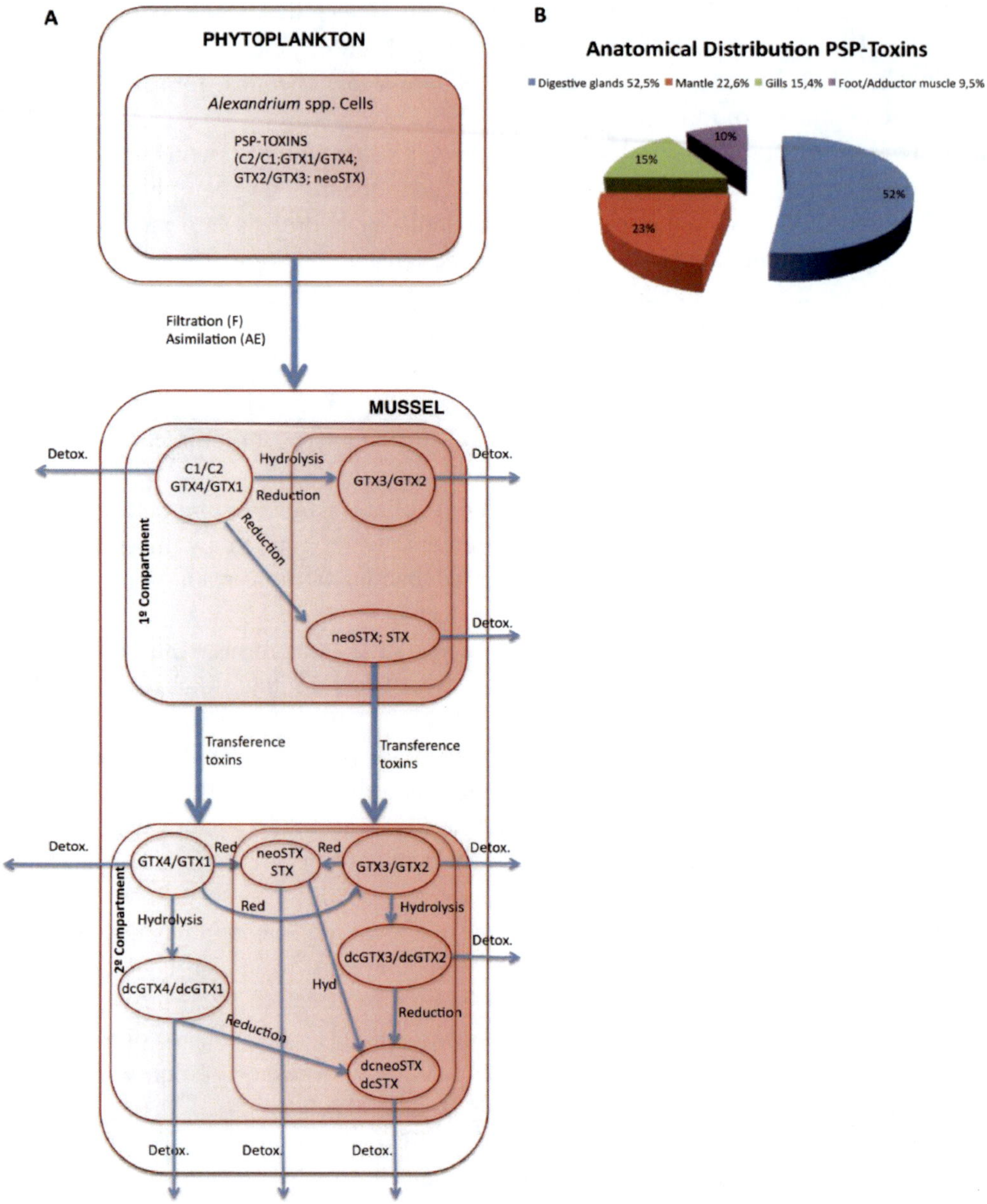

Figure 4. Schematic representation of the distribution, accumulation and biotransformation of PSP-toxins from phytoplankton to mussels' tissues (considering the data provided by Gueguen et al., 2011 [44]). B.- Anatomical distribution (%) PSP-toxins in visceral and non-visceral tissue in mussels.

Therefore, the profile of *Tagelus dombeii* may explain the profile result because it is mainly composed by toxins produced via enzymatic conversion of reductive transformations, generating more stable toxic forms within the profile, such as GTX3/GTX2 (21%) (ratio 0.8:1) with predominance of α-epimer, neoSTX (7%) and STX (72 %) (Figure 3).

Thus, soft bottom congeners (*Venus antiqua or Tagelus dombeii*) are characterized by presenting more diverse profiles, basically determined because their populations predominate in shallow areas (2-3 m deep) and therefore, they are exposed to conditions of increased bloom mobility of toxic dinoflagellates in the water column, which allow them not only to be fed of wastes from the water column, but also directly from endemic toxic dinoflagellates (*Alexandrium catenella*) since the profile is composed by pathways of toxins already subjected to other bivalves' (mussels) metabolic transformations or from algae toxins. An additional fact is the presence of paralytic toxins in gastropod carnivorous in habitats composed by mussels where these carnivorous (*Concholepas concholepas*) are characterized according to the diet and its location in the marine habitat.

Since these species predominate at the rocky intertidal zone where their food is available -*Mytilus chilensis* (1,138 ± 54 μg STX-equivalent 100 g^{-1}) and *Tagelus dombeii* (2,505 ± 101 μg STX-equivalent 100 g^{-1})- from the subtidal zone (sandbanks), it allows defining that their profile is derived from the species of which their diet was based on, being this toxic contribution subjected to a high and effective enzymatic metabolism that generates efficient transformations of toxins from GTX4/GTX1 into GTX3/GTX2 and quickly into STX, thus generating more stable toxic profiles (STX).

Finally, the determination of the toxic profile in mussels is very important, since it allows obtaining information regarding the transfer of toxins and thus determine the degree of conversion during digestion [26, 27, 54, 57-59].

3. Lipophilic Toxins

Lipophilic toxins are formed by a group of toxins of lipophilic nature corresponding to Diarrhetic Shellfish Poisoning (DSP), Pectenotoxins (PTXs) Yessotoxins (YTXs) and Azaspiracid (AZAs) (Figure 5).

Initially, PTXs, YTXs and AZA had been classified within the DSP group, since in a large percentage, during evaluation processes in mussels and phytoplankton, this group of toxins was co-extracted and the symptoms detected in the bioassays or those caused by intoxications of people tended to be associated to the whole group of described toxins. Subsequently, it was determined that this group was composed by different types of toxins, very different from each other from the chemical point of view and with different toxicological properties, so the PTXs, YTXs and AZAs group were excluded from the DSP group and classified as independent groups [10, 60, 61].

The EU has established guidelines for maximum toxin levels in shellfish for human consumption of OA, DTXs and PTXs, whether combined or single value whose edible parts of seafood cannot exceed 160 μg OA equivalent/(shellfish kg), while maximum levels of YTX cannot exceed 1 mg YTX eq./(shellfish kg) and 160 μg AZA equivalent/(shellfish kg) for AZAs.

As of July 2011, the Commission Regulation (EU) Nº 15/2011, amended Regulation (EC) No. 2074/2005, which recognised testing methods for detecting marine biotoxins in live bivalve molluscs and established the EU-RL LC-MS/MS method as the reference method for detection of lipophilic toxins and used as matter of routine, both for the purposes of official controls at any stage of the food chain and own-checks by food business operators [13].

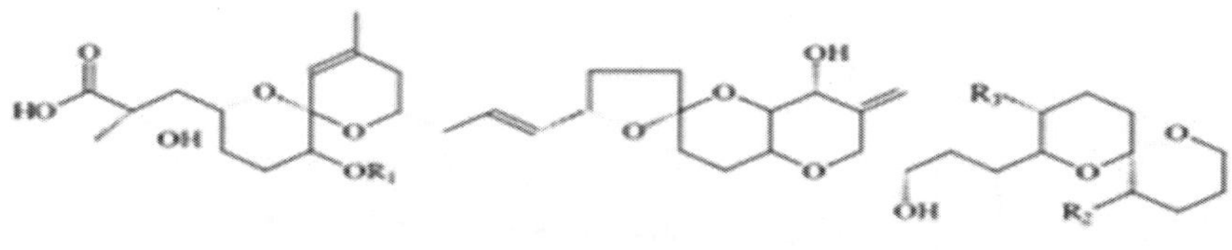

Palmitic Acid

Okadaic Acid	(OA)	R1=H	R2=H	R3=CH3
Dinophysistoxin-1	(DTX-1)	R1=H	R2=CH3	R3=CH3
Dinophysistoxin-2	(DTX-2)	R1=H	R2=CH3	R3=H
Dinophysistoxin-3	(DTX-3)	R1=Acyl	R2=CH3	R3=CH3

a

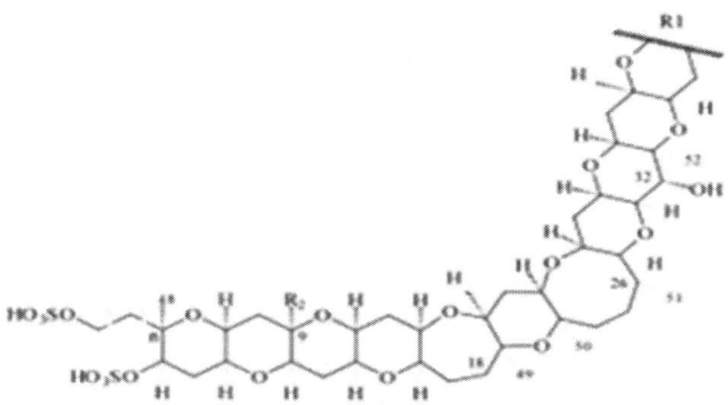

Name	R^1	R^2	n
Yessotoxin (YTX)	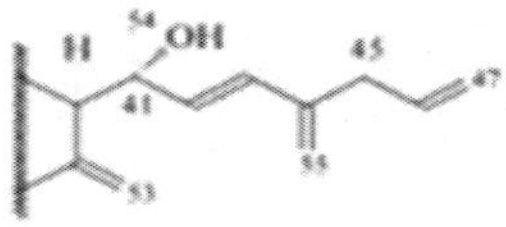	H	1
45-Hydroxy-YTX (45-OH-YTX)	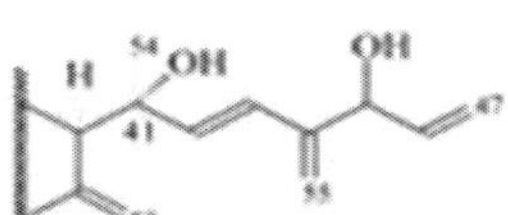	H	1
Homo-YTX (1a-Homo-YTX)	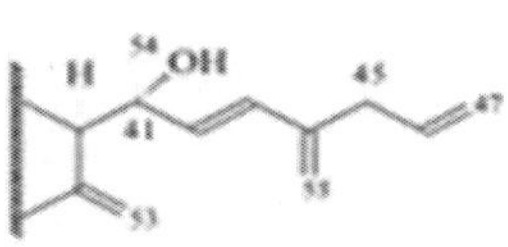	H	2

b

Figure 5. (Continued).

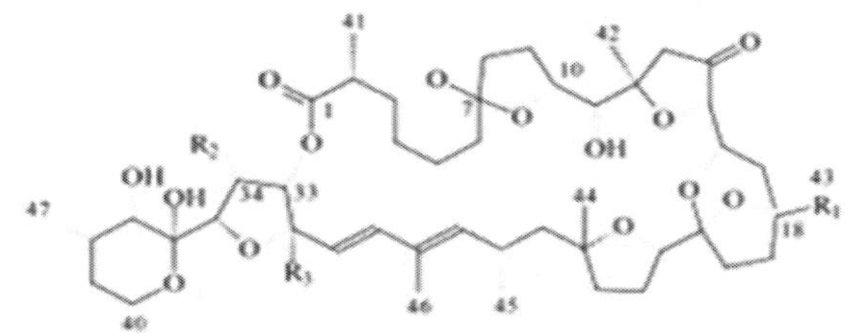

Name	R^1	R^2	R^3	C-7
PTX-1	CH_2OH	H	H	R
PTX-2	CH_3	H	H	R
PTX-2b	CH_3	H	H	S
PTX-3	CHO	H	H	R
PTX-4	CH_2OH	H	H	S
PTX-6	COOH	H	H	R
PTX-7	COOH	H	H	S
PTX-11	CH_3	OH	H	R
PTX-11b	CH_3	OH	H	S
PTX-13	CH_3	H	OH	R

c

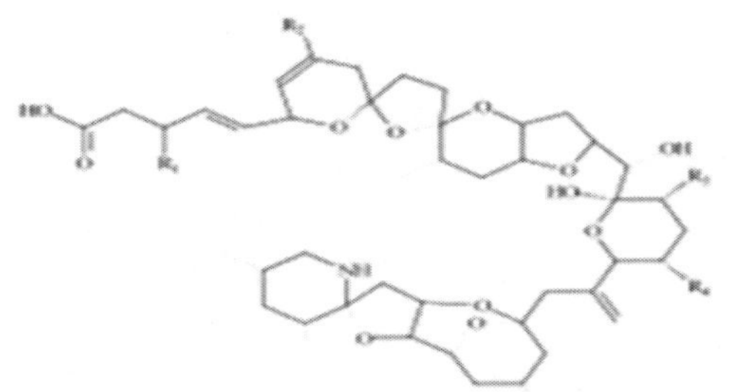

	R1	R2	R3	R4
AZA1	H	H	CH_3	H
AZA2	H	CH_3	CH_3	H
AZA3	H	H	H	H
AZA4	OH	H	H	H
AZA5	H	H	H	OH
AZA6	H	CH_3	H	H

d

Figure 5. Chemical Structures of lipophilic marine toxins. A. Diarrhetic Shellfish Toxins (DSP-toxins); B. Yessotoxins (YTXs); C. Pectenotoxins (PTXs); D. Azaspiracids (AZAs).

3. A. Diarrhetic Shellfish Poisoning-Toxins

Diarrhetic Shellfish Poisoning (DSP) is a gastrointestinal disease caused by ingestion of shellfish with accumulated fat-soluble polyether toxins produced by dinoflagellates of *Dinophysis* and *Prorocentrum* genera. The density of cells associated with a bloom of DSP corresponds to $< 10^4$ cell L^{-1}, with an abundance of $< 10^3$ cell L^{-1}. Intoxications associated to DSP are a worldwide threat to public health and the shellfish industry [7, 62, 63].

The first documented intoxications with DSP-toxins, were recorded in Holland (1961) [64]. To date, high levels of these toxins have been found reiteratively off the coast of Europe

(U.K., Ireland, Denmark, Sweden, Norway, France, Spain, Italy, Portugal, Holland and Belgium); Canada; South America (Chile and Argentina); Japan; Australia and Africa (Morocco) [10,65,66]. In the South Pacific Ocean, blooms associated to DSP usually occur in spring-summer periods (October-March) [67].

Okadaic acid (OA) and its isomers Dinophysistoxin-1 (DTX-1) and Dinophysistoxin -2 (DTX-2) are the main toxins produced by DSP- toxin-producing dinoflagellates (Figure 5A) [68], which present specific toxicity equivalence factors (TEFs), OA= 1.0 (TEF); DTX-1 = 1.58 (TEF); DTX-2 = 0.48 (TEF) [69].

The minimum dose of OA and DTX-1 necessary to induce toxic symptoms in humans has been estimated at 40 and 36 μg, respectively [70]. Death of intoxicated patients has not been reported, but Okadaic acid (OA) and Methyl-Okadaic acid (DTX-1) have been shown to be potent tumor promoters in animals [71,72]. Therefore, they might increase the risk of cancer among regular consumers [73,74].

Toxins associated to DSP are powerful cytotoxins that inhibit various types of serine/threonine phosphoprotein phosphatases 2A (PP2A) [75,76]. These proteins are a group of enzymes that carry out dephosphorylation of numerous proteins, a very important function which is closely related to many essential metabolic processes of eukaryotic cells [77]. OA and DTXs are more potent against PP2A than PP1 (OA-IC_{50} 2.87 ng/ml and DTX-2-IC_{50} 5.96 ng/ml) [75,76].

Depending on the species and natural depuration, DSP toxins are modified inside the molluscs, where OA, DTX-2 or DTX-1 are esterified in the hydroxyl group of Carbon 7, being palmitic acid the most common fatty acid found in Dinophysistoxin-3 (DTX-3) (Figure 5A). Normally, palmitic acid amounts to 90% of the total fatty acids linked to the 7-*O*-acyl-derivative of Dinophysistoxin-1 to produce DTX-3 in chilean samples [15], although other forms of acyl derivatives, such as C14:0, C16:1, C16:0, C18:1 and C18:0 have also been detected in *M. galloprovincialis* and *D. trunculus* in samples collected at the Portuguese coast [78].

Vectors play an important role in transmitting diarrhetic shellfish toxins (DST) through the food chain, causing unknown toxic effects to other organisms [79-81].

OA periodic exposure in mussels results in changes in gene expression, by specific interactions of DST with receptors or indirectly by inducing intracellular signaling cascades, producing loss of stabilization of focal adhesion and cytoskeletal disorganization [82,83].

Since mussels are the main vector of DSP and its contamination causes great harm to mussels coming from natural beds and/or from farming facilities, multiple methods of detoxification, which require to be quick, efficient, easy to apply and may not alter the organoleptic qualities of products, have been developed. Among these processes, thermal shock, freezing, evisceration, supercritical CO_2 with acetic acid, ozonation and radiation are highlighted; these processes are characterized by destroying or reducing the toxicity of mussels [84].

From the above mentioned processes, the most widely used corresponds to radiation, because it is able to destroy pests and pathogens in large production processes of different kinds of foods [85,86]. Thus, mussels containing toxins close to the regulatory limit of 160 μg OA equivalents kg^{-1} mussels' tissue, when subjected to γ - irradiation, show a marked decrease in their toxic content, resulting in products suitable for consumption and sale. However, it should also be considered that this procedure can lead to highly toxic or carcinogenic compounds (2-alkylcyclobutanones and radyolitic derivatives of triglycerides) [87].

Another highly used method corresponds to ozonation; when this method is applied to homogenized mussels, it generates a highly positive and favorable effect for the destruction of DST, this is attributable to the penetration capacity of the gas (ozone) on the mussels' tissue, thus, having the option of directly interacting on the molecules of DST. However, when the toxic profiles of mussels are dominated by toxic varieties such as acyl-derivatives (acyl-OA or DTX-3), the ozonation process, may cause a decrease in ester forms (DTX-3), producing a DST increase in nonacylated forms (OA or DTX-1). Therefore, ozonation may alter the toxin profile of DST toxins, being a final product (canned mussels) happen to be dominated by highly toxic forms of DST, considering besides that the toxicities of 7-O-acyl-esters forms may have similar toxicities than their DSP-toxins if esters are hydrolyzed in the human digestive system [88].

From the toxicological point of view, the toxicity of mussels highly depends on the variability of lipid content in hepatopancreas (≈ 10 - 20%), which favours the toxins' retention factor [67, 89, 90]. That explains why these toxins mostly predominate in this tissue, in comparisson with the whole body which has a lipid content of only 3%.

Despite the high levels of exposure to blooms associated by DST, mussels do not show mortality effects of species from natural beds or culturing centers during their adulthood and this may be attributed to a protective effect of mussels against blooms. *In vitro* studies have established that PP1 and PP2A from mussels are inhibited by DST (OA and DTX-1) like those from any organism. However, *in vivo* studies have demonstrated that toxic effects of DST on cytosolic enzyme (glycogen synthase) are lower, suggesting that mussels have a mechanism that prevents the accumulation of DST in ambient cytosolic. This response against the intracellular effects of OA is mainly due to multixenobiotic resistance (MXR) in mussels blood cells, which by being exposed to OA are more resistant against the toxin's cytotoxic effects [91]. This route of resistance is related to the over-expression of a plasma membrane phosphoglycoprotein (p-gp) which is preferably in gill, mantle tissue, digestive glands membranes vesicles and blood cells and which functions as an ATP-dependent drugs effux pump [92, 93].

Another protective mechanism associated to mussels exposed to DST is related to the uptake and storage of OA/DTX-1 in the lysosomal system and whose organelles are known by concentrating foreign compounds, such as lipophilic xenobiotics. Thus, contaminants such as OA/DTX-1 would be captured and carried through vesicles to lysosomes, predominantly favoring it, because toxins filtered by mussels (OA/DTX-1) are associated with degraded algae fragments, so they are captured by endocytosis to intracellular digestion in lysosomals compartments, which would finally explain why there is no accumulation of DST (OA/DTX-1) in cytosolic in mussels and, therefore, no effect on the activity of cytosolic enzyme regulated by PP1 and PP2A [89].

Since DSP is a serious problem for aquaculture worldwide, it has been established that its predominance in the aquatic environment further responds to the ability that toxins which compose them have to inhibit the growth of a variety of microalgae in a concentration-dependent manner, thereby suggesting an allelopathic role of toxins. In dinoflagellates, cells mainly produce toxic forms such as OA-diol-ester and Dinophysistoxin-4 (DTX-4) which are rapidly hydrolyzed to produce OA or DTX-1. In the early stages of production of toxins, DTX-4 is the majority toxin and, therefore, the route of excretion of toxins from cells, which is rapidly hydrolyzed to OA-diol-ester and then to OA by enzymes, and it may even be effective against predators that break their cells during feeding. Through *in vitro* studies, it has been determined that DTX-4 is 50 times less effective than OA on PP2A inhibition, where the OA-diol- ester toxic form is relatively inactive [94, 95].

An important factor is related to the fact that since OA is a lipid-soluble polyether, its carboxylic group can be ionized at physiological pH, which would make diffusion difficult to it through the membrane. Thus, since pre-toxic forms such as OA-diol-ester have no ionizable groups, they can move through the membrane with relative ease, allowing it to mobilize into the cell and being inwardly hydrolyzed acquiring a toxicity of 1,000 times higher than its predecessor.

3. B. Yessotoxins Group

Yessotoxins group (YTXs group) are sulphated polyethers (Figure 5B). They were isolated for the first time from the Japanese scallop *Patinopecten yessoensis*, and have also been detected in mussels' samples in Norway, Italy, New Zealand, Chile, Russia, Canada, United Kingdom and España [60, 96-103]. They have been identified in both, shellfish and potentially causative dinoflagellates: *Protoceratium reticulatum, Lingulodinium polyedra*, and *Gonyaulax spinifera* [98,104,105]. Blooms associated to YTX-producing dinoflagellates are characterized by a concentration of cells of 10^3 cells L^{-1}, where the production of YTX toxins in dinoflagellates corresponds to an average of 34 pg $cell^{-1}$, in which homo-YTX and YTX are the toxins that mainly compose the profile in variable concentrations and depending on the area where they had been identified [60].

Historically, YTXs have been classified within the DSP-toxins group, however, these toxins do not induce diarrhea in mice bioassays, nor produce protein phosphatase inhibition, reason enough to classify them in a separate group [106,107]. The precise mode of action remains currently unknown, although YTX is known as a potent phosphodiesterase activator [108].

Symptoms caused by poisoning with YTX in humans are relatively unknown because any human intoxication case has been reported to this date [61]. YTX toxic evaluation, in bioassays, has determined that after an intraperitoneal injection of approximate concentrations of 150 μg/kg, it generates dispersed, mobility dispersed, dyspnea, tremor and cramps, symptoms that begin 4 hours after the injection was given [108,109]. EFSA has established that toxic effects may not occur if the concentrations do not exceed the levels of 3.75 mg YTX-equivalent Kg^{-1} shellfish [10, 110].

Currently, 100 natural derivatives of YTX have been identified and characterized by NMR and detected by liquid chromatography coupled with mass spectrometry (LC-MS). Some YTXs are directly produced by dinoflagellates (norYTX, 41-keto-YTX and 41a-homo-YTX) while shellfish metabolizes through different oxidation pathways, such as hydroxylations, carboxylations, desulfations, methylations and amidation to other YTX analogs [60,97,111].

The main vector of YTXs corresponds to scallops and mussels, which can accumulate a large amount of toxins in their tissues due to the high filtration capacity they have. Hepatopancreas is the main tissue of accumulation and according to chemical modifications caused by mussels, they can spread to other tissues such as mantle, gill and adductor muscle

[112]. Thus, YTX-toxins have been found in gastric glands of mussels, associated to several enzymatic metabolism derived toxic forms, of which new forms have been identified so far such as 45-hydroxyYTX, 1a-homoYTX, 45-hydroxy-YTX and 45-hydroxy-1a-homoYTX (Figure 5B) [113]. Each of them present a specific toxicity equivalence factor (TEF), YTX= 1.0 (TEF); 1a-homo-YTX = 1.0 (TEF); 45-OHYTX = 1.0 (TEF) and 45-OH-1a-homo-YTX = 0.5 (TEF) [110].

When mussels are exposed to this type of bloom composed by yessotoxin-producing dinoflagellates, they absorb algae and the YTX-toxic content is rapidly oxidized via enzymatic action- to 45-OHYTX to be then slowly transformed to carboxy-YTX, isomer which is more characteristic in the depuration phase of mussels [114-116].

3. C. Pectenotoxins Group

Pectenotoxins (PTXs) are a family of polyether macrolide toxins (Figure 5C) which are produced by the same species that produce toxic forms of DSP, namely: *Dinophysis fortii*, *Dinophysis acuta*, *Dinophysis acuminata*, *Dinophysis caudata* and *Dinophysis norvegica* and in heterotrophic dinoflagellates such as *Protoperodinium divergens*, *Protoperodinium depressum* and *Protoperodinium crassipes* [117,118]. Their names come from the name of the shellfish were each of them was first found, for example, Pectenotoxin-2 (PTX-2) was isolated from the Japanese scallop *Patinopecten yessoensins* [119]. To date, however, numerous PTX analogs have been found in algal cultures, shellfish and blooms associated to dinoflagellates that produce this kind of toxins.

Several new isomeric forms of PTX-2, such as PTX-4, PTX-6 and PTX-7, have been found in natural blooms (Figure 5C). Also, when filtered by shellfish and metabolized in the hepatopancreas (digestive glands) they undergo biotransformation. PTXs have been reported in Ireland, Croatia, New Zealand, Portugal, Norway, Japan and Chile [61,67,120]. To differentiate isomeric forms of PTX-2 found in dinoflagellates from those found in shellfish, the nomenclature coined the term PTX-seco acid (PTX-SA) [61]. Toxic isoforms such as Pectenotoxins-2 (PTX- 2), Pectenotoxins-2 seco acid (PTX2sa) and 7-epi-Pectenotoxins-2 seco acid (7-epi PTX2sa) are the predominant toxic forms found in Europe [121]. To date, 15 forms of toxic analogs of PTX have been described, the last identified isomers esterified forms of PTX-2A-SA (37-O-acyl-ester-PTX-2-SA, 11-O-acyl-ester-PTX-2-SA and 33-O-acyl-ester-PTX-2-SA) found only in shellfish but not in PTX producing dinoflagellates [111, 122].

The coextractability of PTXs with diarrhetic toxins of okadaic acid group led to PTXs to be included with okadaic acid analogues in the diarrhetic shellfish poison toxins group [123, 124]. However, PTXs are now classified as a separate toxin group, on the basis of their toxicity, thus for example, PTX1 and PTX6 do not inhibit PP2A. Consequently, based on the associativity of symptoms and the toxic ability of mussels, it has been determined that the amount sufficient to cause toxic symptoms is 2.0 μg PTX-2-SA/Kg for a 70 kg person [125].

Available information on PTXs is still insufficient with just a few biological results. These few results are about *in vitro* and *in vivo* experiments, aiming to provide information about the risks to human health that ingestion of these toxins may pose [61]. PTX-2 toxicity after intraperitoneal (ip) administration and after oral ingestion is considered comparable. For example, hepatic damage in mice after ip administration and oral ingestion seem to be the same.

However, diarrhea was never reported in those experiments [118]. In addition, PTX-2 has been found to be potentially citotoxic in lung, colon and kidney carcinogenic cell lines, an effect that cannot necessarily be extrapolated to its toxic forms, such as PTX-2-SA and 7-epi-PTX2-SA, evidencing the importance of the initial structure of toxins that initiate citotoxic effects [126], where 7-epi-PTX-2-SA is the result of interconversion of PTX-2-SA to thermodynamically more stable forms and may be understood as a detoxification protective effect by mussels [118, 127].

To date, no case of intoxication by PTX-type toxins has been reported, although biotransformation of PTX-2 has been found in some shellfish species and these toxic forms have been detected with increasing frequency [118].

Studies from various toxins have determined a difference of profiles from producing dinoflagellates and bivalves'profiles, establishing that the metabolic transformation occurs in mussels after ingestion of toxic algae [113]. In algae, the identified majority toxin only corresponds to PTX-2, which once it is accumulated in the digestive glands of mussels and via oxidation of 43-methyl-group generates PTX-6.

3. D. Azaspiracid Toxins Group

Azaspiracids toxins group (AZAs) are produced by *Protoperidinium crassipes* and *Azadinium spinosum* species [128,129], although direct evidence relating between the high level of cells that produce AZAs and the concentration of toxins in shellfish that filter them, has not been yet provided [130]. The first intoxication cases related to this type of toxins happened in Holland in 1965, where the symptoms were initially linked to the presence of OA and DTXs.

To this date, 24 different forms of Azaspiracids (AZAs) have been found, being Azaspiracid-1 (AZA-1), Azaspiracid-2 (AZA-2) and Azaspiracid-3 (AZA-3) the most common forms (Figure 5D) [131]. All of them are characterized by having specific toxicity equivalence factors (TEFs) (AZA-1 = 1; AZA-2 = 1.82; AZA-3 = 1.43) [132, 133].

Identification of AZAs in shellfish has been reported in Ireland, UK, Norway, France, Portugal, North of Africa (Morocco), Chile and U.S.A. [134-136]. However, new forms of AZAs are detected every year by the monitoring programs in countries of the EU, which translates to variable toxic profiles, depending on the species that accumulate the toxins and the extent of their biotransformation in the hepatopancreas. These last isoforms of AZAs, have only been observed with LC-MS/MS analysis, which has been implemented as an alternative technique in sanitary shellfish controls [137, 138].

Action mechanisms of this type of toxins has not been clearly determined, although preliminary in vitro studies in mammal cells show alterations in the cytoskeleton structure and on the E-cadherin system, responsible for cell to cell interaction [139, 140]. These experimental results could somehow explain symptoms associated to the described intoxications; gastrointestinal disorders, abdominal pain and diarrhea [141], since bioassays in mice with shellfish extracts with low concentrations of AZA do not induce diarrhea [142], however after a while animals show paralysis and respiratory disorders leading to dead about 1.5 hour post injection of the extract [132]. Further studies have demonstrated that lethal i.p.

dose in mice depends on the toxic form present in the shellfish extracts. AZA-1 has a lethal dose of 200 μg/kg, while isomers AZA-2 and AZA-3 are significantly more toxic with lethal doses of 110 μg/kg and 140 μg/kg, respectively [133].

Biotransformation may play a significant role on the number of toxic variations within a shellfish. Toxic variations AZA-4 and AZA-5 were found in mussels that originally only had AZA-2 and AZA-3 (Figure 5D) [143, 144]. These biotransformed variations, although less toxic, constitute a source of concern from the public health standpoint [135, 145].

On other hand, oral toxicity of AZA-1 at high doses (900 μg/kg) in mice, produce significant damage to the small intestine, while 500 μg/kg dose produce only hepatic damage, evidenced by a volume increase of the organ (38% post-ingesta increase) [146]. Parallel toxicological studies have established that the lowest observed effect level (LOEL) for AZA lies between 23 and 86 μg/kg by person, even when levels of 80 μg AZA eq/kg of shellfish could not produce AZA related intoxication symptoms. However, doses of 30 μg AZA eq/kg (shellfish kg) in 400 g shellfish ingested have produced intoxication syndromes in humans. These seemingly contradictory data can be reconciled by appealing to biotransformation underwent by AZA, producing different toxic forms (isomers), during their stay inside the shellfish [147]. Based on both the toxic information in humans and the biotransformations determined in mussels, EFSA has established that toxic effects in humans may not occur if the concentrations of AZAs do not exceed the levels of 30 μg AZA-1-equivalent Kg^{-1} shellfish [110]. Proteomic studies on hepatopancreas of *Mytilus edulis* exposed to AZA, indicate that they show an increase in the protein levels associated with the defense mechanisms of mussels, thus establishing that hepatopancreas plays an important role in the metabolism of toxins [148]. Toxicity in mussels contaminated with AZA persists for an average of five months, being in some cases mainly distributed in meat-compartments more than in the digestive glands [141]. Even when AZAs are usually concentrated in hepatopancreas of mussels, these toxins may migrate to other tissues of mussels (gonads and adductor muscle) given their chemical qualities of zwitterion [141, 149].

Under production conditions, mussels contaminated with AZA-1 which are subjected to short-time thermal processes and in the absence of water, experience high levels of AZA-3, a situation that is repeated by subjecting mussels at low temperatures (4°C) for prolonged times. In both instancies, the toxic precursors AZA-1 and AZA-2 have no considerable variations to the originally evaluated levels, even when the levels of AZA-3 correspond directly to those provided by AZA1/AZA-2. A plausible answer is that toxic contributions come from other sectors of the matrix of mussels or also by bioconversion from unidentified toxins such as AZA-17, which is decarboxyled when subjected to high temperatures, generating AZA-3. A similar situation occurs with decarboxylation of AZA-19, AZA-21 and AZA-23, which tend to form AZA-6 [150]. Important factors which are emphasized when considering that multiple production processes of mussels may damage the walls of their internal bodies in order to facilitate the distribution of toxins to other compartments.

3. E. Composition and Biotransformacion of Lipophilic-Toxins in Mussels

Assessment and identification of lipophilic toxins (DSP-toxins, YTX, PTX and AZA) and of their described and identified toxic forms resulting from metabolism processes (acyl-derivatives) in shellfish, whether these come from rocky strata or soft bottoms habitats, evidence that the profiles are always characterized by the predominance of DTX-1 (65% average among species), followed by a toxic variability dependent as to the involved species. Hence, the sequence of predominance of toxins in *Mytilus chilensis* is composed by DTX-1>OA>PTX-2>YTX>AZA-1, where each toxin is followed by the toxic forms which have resulted from the metabolic transformations occurred in the mussels's tissues (Acyl-DTX-1>PTX-2SA>45-OHYTX>7-epi-PTX-2SA>homoYTX>AZA-2>Acyl-OA>AZA-3). For its part, Ribbed mussels (*Aulacomya ater*) are characterized for having a profile dominated by DTX-1>PTX-2>YTX>AZA-1>OA and their toxins resulting from biotransformation by PTX-2SA>Acyl-DTX-1>45-OHYTX>7-epi-PTX-2SA>Acyl-OA>homoYTX>AZA-2>AZA-3 (Figure 6). Meanwhile, the profile of dominance of toxins in bivalves in soft bottom species (sand), which are part of the infauna, are likewise different. The dominance criteria for *Tagelus dombeii* is DTX-1>YTX>PTX-2>OA>AZA-1, where Acyl-DTX-1>45-OHYTX>AZA-2>homoYTX>AZA-3>Acyl-OA>PTX-2SA >7-epi-PTX-2SA dominate in its biotransformed toxins (Figure 6). The main toxic forms of the biotransformation of PTX-2 correspond to PTX-2-SA/7-epi- PTX-2-SA, which may dominate in the variety of PTX in mussels after one month. This dominance is related to the chemical stability of metabolic forms and to the high retention affinity in the mussels' tissues, relating also this transformation pathway to protective effects towards mussels. This toxic profile may present strong toxic stratification in mussels from farming facilities, because the availability and assimilative capacity of food may vary as the depth increases or decreases. The identification of AZA-1 in mussels was dominant and also directly related to their interaction with the other aquatic organisms in the related habitat.

The presence of AZA-2/AZA-3 isomers, known toxins structures that result from the bioconversion of AZA-1 in mussels is also established. (Figure 6) The variability in the accumulation of lipophilic toxins in shellfish can be largely explained by the different degrees of hydrophobicity that toxins have in the different species and besides it must be considered their direct relationship in the lipid content, where Blue mussels (*Mytilus chilensis*) and Ribbed mussels (*Aulacomya ater*) are dominant species, known by having high and varied concentrations of lipids [83, 90, 151-153]. The detection of toxic forms resulting from biotransformation, evidences the conversion capability that mussels and habitat-associated species have to modify lipophilic toxins, where the transformation ratio is again associated with the degree of compartmentalization of toxins provided in the bloom and where each of them will be chemically modified according to the degree of exposure of the species to the bloom and also to the degree of retention or purification by the aquatic organism. Such factors may also vary according to the availability of food by filter-feeding organisms, where the diet may be also subjected to food from faeces and pseudofaeces more directly exposed to the water column, producing a contribution of toxins which were already subjected to metabolic or biotransformation processes by other species present in the water column (Figure 7A) [80, 111, 154].

The main accumulation's tissues of lipophilic toxins in mussels corresponded to digestive glands (hepatopancreas), in a ratio of 7:3 with respect to non visceral tissues (gonads, gills, mantle and adductor muscle) (Figure 7B-E).

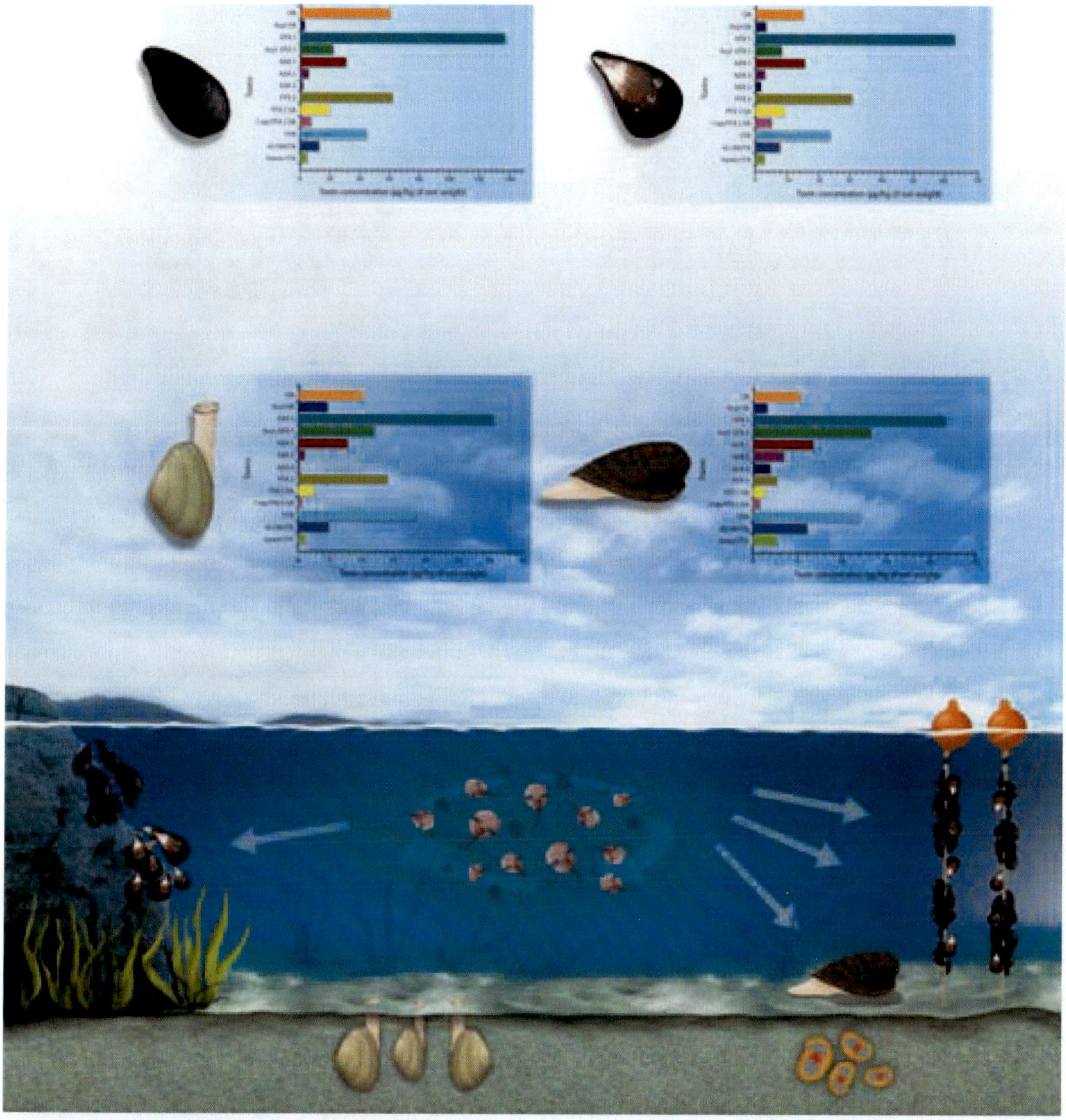

Figure 6. Comparison of lipophilic toxins profiles of bivalve species Blue mussels (*Mytilus chilensis*) Ribbed mussels (*Aulacomya ater*), Razor Clam (*Tagelus dombeii*) and carnivorous gastropods Loco (*Concholepas concholepas*) from the South Pacific Ocean.

This distribution is variable in time, tending to change the profile of predominance of toxins in each tissue, given by the biotransformation that toxins undergo, the affinity of toxins in the mussels'compartments and by the different removal rates of toxins.

The transformation of DTX-1 by mussels favors the accumulation of DTX-3 preferently in mantles, which can reach an accumulation level of 90% depending on the period of evaluation after the bloom ($\approx$ 5 months).

Toxic profiles of carnivorous species such *Concholepas concholepas* were characterized by having profiles directly related to the typical diet (*Mytilus chilensis*) and by the high biotransformation capacity they have.

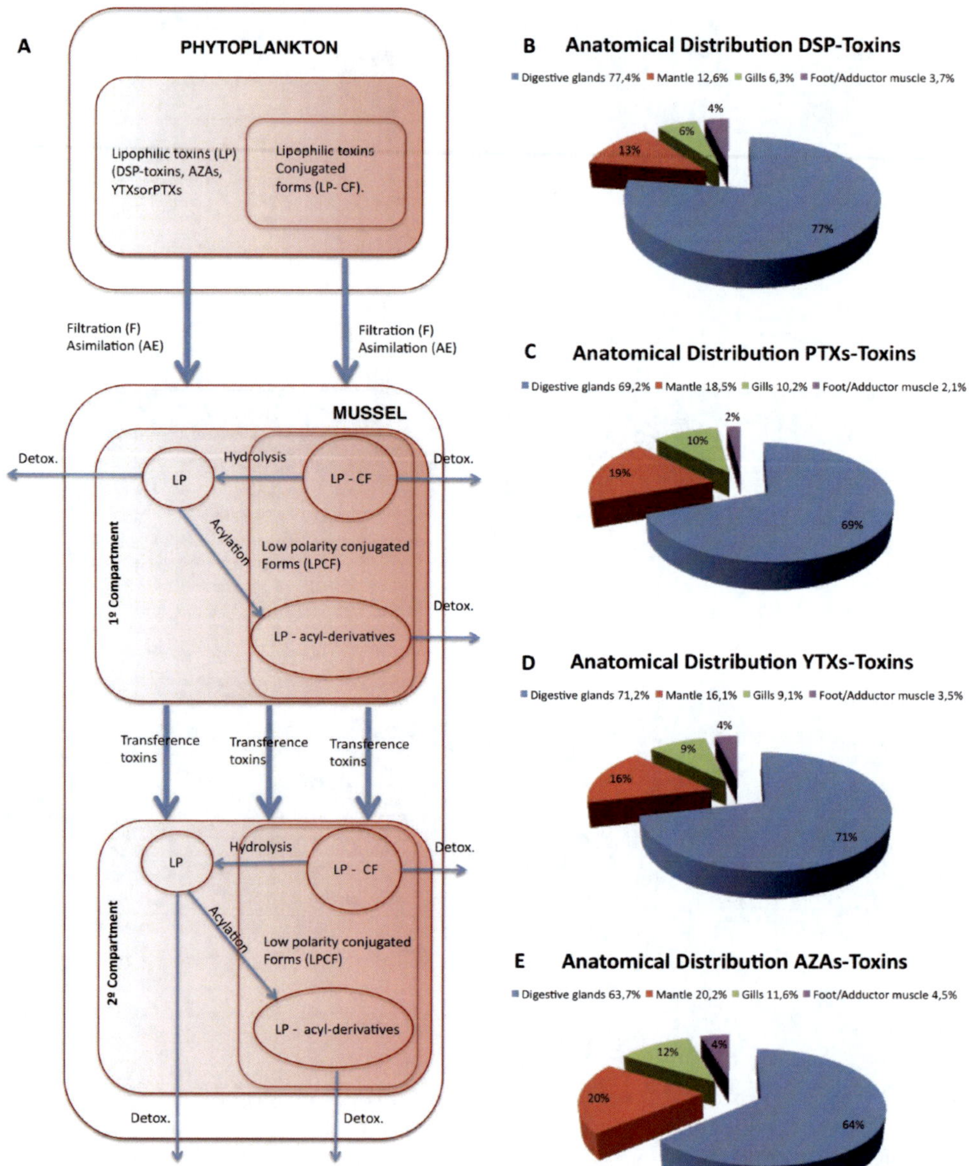

Figure 7. Schematic representation of the distribution, accumulation and biotransformation of lipophilic-toxins from phytoplankton to mussels' tissues (modified and updated of Moroño et al., 2003 [162]). B.- Anatomical distribution (%) DSP-toxins in visceral and non-visceral tissue in mussels; C.- Anatomical distribution (%) PTXs-toxins in visceral and non-visceral tissue in mussels; D.- Anatomical distribution (%) YTXs-toxins in visceral and non-visceral tissue in mussels; E.- Anatomical distribution (%) AZAs-toxins in visceral and non-visceral tissue in mussels.

The precise mechanisms of purification of these kind of toxins in the assessed species are unknown, although the literature suggests a relationship between quality and quantity of available food, together with additional elements such as salinity and temperature conditions, factors that could affect the purification levels of bivalves, and which periods may last up as long as 44 days after the bloom occurred [83, 155].

However, parameters regarding the chemical configuration of some toxins and their purification levels in bivalve species, establish that esterified forms from OA, DTX-1 and DTX-2 toxic forms are purified faster than free forms (nonesterified forms), depending largely on the involved fatty acid, which directly influences the rate of purification of toxins (≈10-17 days) [80].

From the point of view of the contribution of lipophilic toxins to different species, variability can be explained on the basis of the predominance in the bloom of dinoflagellates species that provide this type of toxins and the profile of toxins provided from cells to bivalves considering, for example, that the cells concentration of *Dinophysis sp.* in the water column may represent a small fraction in it, since according to the oceanographic conditions, this species is able to experience vertical migration generating growth according to the nutritional needs [156].

The lattest is a no less important point, considering that according to the abundance of food, in this case of toxic dinoflagellates, the filtration rate of the different species of bivalves may experience a decrease in their physiological needs for food on a regular basis [157].

Furthermore, the determination of the levels of lipophilic toxins has determined the different retention levels that bivalves and gastropods have and that, similarly, some species have a higher retention capacity or a greater degree of insensitivity to these toxins, because when they are subjected to repeated blooms, early stages of development of some species could be affected due to the low effectiveness of their detoxification enzymatic processes, thus decreasing bivalves or gastropods populations.

It should be pointed out that these blooms may affect various larval stages of the different studied species (mussels) by affecting the oxygen uptake rate by larvae, which causes a reduction in their capacity for locomotion.

This generates a negative effect since it directly affects the migration of different larval stages involved in vertical migration in the water column, making that the different larval stages may be exposed to considerable densities of predators (chaetognaths, fishes and/or hydrozoans) affecting the balance of the adult population in a zone of great impact on economically profitable resources and with some endemic cases in the country.

4. Impact of PSP-Toxins and Lipophilic Toxins on Aquaculture Development in the South Pacific Fjords

The *Mytilus chilensis* samples collected in the area are exclusively destined for aquaculture development in the South Pacific Ocean (Chiloe Island and Austral fjords) and given the constant conditions of annual blooms, these tend to occur in summer periods (January – April), where mussels tend to present variable levels of PSP-toxins and lipophilic toxins, which may exceed international standards. The different amounts of each type of detected toxins are associated with the location of the sample in the water column, since each sample is collected at different depths of the shellfish aquaculture lantern. Mussels that are at lowest depth (10 meters) and in the most superficial positions (1-3 meters from the surface) show the lowest amount of DSP toxins, 67.2 and 33.5 – 36.7 nanograms of DTX-1 / gram of digestive gland, respectively. Similar results are obtained in PSP toxin analysis. At a depth of

10 meters, the shellfish PSP toxin content is approximately 87.9 μg / 100 g of mussel meat and at 1-3 meters is 78.3 and 98.7 μg of PSP toxins/100 g of mussel meat, respectively.

These data show a clear stratification of the two toxins (PSP and DSP toxins) in the shellfish samples in the water column. This stratification measured in the mussel extracts obtained from mussels from aquaculture at different water column depths can only be explained by the stratification of the toxic dinoflagellates in the water column.

A not less important factor if it is considered that the exploitation of these resources generates production batches that may come from different areas of the same culturing area and also from different depths, which in a first instance may cause errors in the determination of the profile and toxicities of the exploited mussels, tending to qualify products as free from PSP-toxins and/or lipophilic-toxins.

These data become relevant, considering that 90% of Blue mussels that Chile exports (from Austral fjords) are farmed mussels, whereas exported Ribbed mussels are harvested from wild banks. The notorious differences in toxin levels found are directly related to the filtration capacity of each species.

In recent years, the production of mussels has been subjected to the determination of lipophilic toxins previously unidentified (AZAs and YTX). This impact, that is more relevant during spring–summer season in this region, may be attributed to the presence of algae, mainly to the favourable luminescence and nutrient conditions during this time of the year. A not less important fact we must consider is the Chaiten Volcano eruption (Chiloé Island surroundings), Aysén Volcano eruption (Austral fjords) and the earthquake–tidal wave in the year 2010 – phenomena that could explain the population dynamics change of marine microalgae and that should allow the development of dinoflagellates that produce different types of toxins (YTX, AZA and PTX) in areas of higher production of mussels from aquaculture.

In addition, the latitudinal distribution of phycotoxins along the coasts of Chile in Austral fjords, in which some studies have clearly detailed the presence of isomers of YTX, DSP-toxins and AZAs has been tried to be defined [158]. However, geographical definitions associated with the presence of lipophilic toxins or with PSP-toxins are an off-shore assessment in certain points but they cannot be considered as a representation of the behavior of HABs in Austral fjords, areas where farming centers of mussels are preferably set up. It has been determined that *Alexandrium catenella* may exhibit preferences for certain sea and habitat conditions, which could explain the regional toxic differences in the fjords of the same species and that result in different initial toxic profiles prior to the consideration of a contribution of toxins' biotransformation by bivalves [65,67,159]. Comprehensive assessments on the complex oceanography and geography of austral fjords show very different behaviors in the presence of HAB, in which local zones (estuaries), that make up the fjord's geography, determine very different levels of *Dinophysis sp.* blooms' behaviors in different points [160]. Therefore, individual (oceanographic) characteristics of each fjord make the HABs' behavior very different and unique, or difficult to characterize, among others, by particular physio-chemical parameters in some cases.

The distribution of different Dinophysis species (*Dinophysis acuta* and *Dinophysis acuminata*) has been defined as variable and it is directly associated to oceanographic conditions, determining the simultaneous presence of species, with a variability associated to the water column. The characterization of a region composed by multiple fjords is delimited to the transitions from upwelling to downwelling in fjords in different times of the year

(Inland seas) [161]. Therefore, defining a characteristic condition of a profile or predominance of *Dinophysis sp.*, *Alexandrium sp.* or a toxin in particular, responds directly to the methodological and sampling inadequacy in space-time with which HABs are related in this zone of austral fjords.

Conclusion

Harmful Algal Blooms (HABs) is a natural phenomenom caused by different types of cells (dinoflagellates, cyanobacteria and diatomae) and is characterized by producing different types of toxins which, through the trophic chain that involves hydrobiological organisms, may affect human health. PSP-toxins (STX-isomers) and Lipophilic-toxins (PSP-toxins, DSP-toxins, YTXs, PTXs and AZAs) correspond to the main group of toxins that contaminate most of the shellfish worldwide, mainly *Mussels sp.* derived from natural beds and those produced through aquaculture development".

Mussels and bivalves exposed to HAB accumulate filtered microalgae in their digestive glands (hepatopancreas) where the 70% of the total amount of the detected toxins is present in this organ. From these glands, toxins are subjected to biotransformation mechanisms by basically three types of pathways:

a) By bacteria such as *Pseudomonas sp.* and *Vibrio sp.* which are an important part of their environment.
b) By aquatic organisms which are characteristic of diverse habitats.
c) By enzymes identified in bivalves and which play an important role in esterifying and metabolizing toxins into chemically more stable, easy to purify isomers by mussels.

The mechanisms of biotransformation of toxins present in bivalves lead to the formation of new pseudo-toxic forms that are directly associated with the stage of excretion of toxins by bivalves where these new pseudo-toxic forms are many times not detected by conventional detection methods (bioassays), and their intraperitoneal and oral toxicities are not related, which involves to generate naturally contaminated products as toxin free products, causing many times that these nontoxic forms are again transformed into toxic forms during production processes of processed seafood or by enzymatic action in the digestive processes of the human body. Hydrolysis and acylation processes in bivalves correspond to widely studied mechanisms, clearly showing that the stages of esterification of toxins by fatty acids catalyzed by enzymatic action correspond to a mechanism of elimination of toxins by some invertebrates, being determined that highly esterified proportions of toxins (conjugated forms) are detected in the faeces of bivalves and that this degree of biotransformation depends largely on the type of toxins which were initially present (hydrophobicity). This lipophilic toxin biotransformation process is relevant because it cannot be detected by conventional monitoring methods (such as mouse bioassay), which then may classify the tested individuals as toxin free bivalves or could lead to an erroneous definition of bloom effect depuration processes.

An additional factor to the variability process of the detected toxins level (DSP-toxins, YTX, AZA and PTX) among the assessed species, corresponds to the stratification of dinoflagellates (microalgae) in the water column and to the filtration capability that the different species have, which explains the variability of the profiles of the different assessed organisms, whether from natural beds or farming facilities. The need to determine the profiles of toxins in contaminated bivalves and their biotransformation levels during or after an exposure to HAB in order to define the levels of toxic and chemical modifications of toxins in bivalves subjected to industrial manufacturing processes, has become highly necessary because of two essential facts:

1. Because of the high variability and transformation probability to which toxins are subjected to enzymatic processes in bivalves which impedes the obtention of adequate information about the evolution of toxic events and/or purification processes in bivalves.
2. Because of the difficulty of identifying new conjugated or biotransformed forms by conventional methods such as bioassays, since there is no reference material available in most cases which impedes the identification and quantification of new toxic forms.

Knowledge of the interactions among contaminated shellfish and the evolution of toxins' transformation in this study may allow performing a resources' efficient risk management, making its extraction more efficient and avoiding adverse effects from the economic point of view due to the extended closures enacted as a result of toxic blooms.

REFERENCES

[1] Asmus, R.M. and Asmus, H. 1991. Mussels beds, limiting or promoting phytoplankton. *J. Exp. Mar. Biol. Ecol.,* 148, 215-232.

[2] Avila, M., Plaza, H., Norambuena, R. and Bustos, E. 1995. Estado de Situación y Perspectivas de la Acuicultura en Chile. Etapa II. IFOP/CORFO. 219 pag.

[3] Cortez, C., Nuñez-Aros, R., Astorga, M., Perez, M., Presa, P., Piña, B., Martínez, V., Toro, M.A. and Toro, J. 2012. Miticultura y Ciencia: Visión desde las empresas. In: Astorga, M., Toro, J. and Martínez, V. (Eds), Genética de Mitílidos y su Impacto en la Miticultura (14-19).Valdivia, Chile. UACH.

[4] Cloern, J.E. 2001. Our evolving conceptual model of the coastal eutrophication problem. *Mar. Ecol. Prog. Ser.,* 210, 223-253.

[5] Newells, R.I.E. 2004. Ecosystem influences of natural and cultivated populations of suspension-feeding bivalve molluscs: A review. *Journal Shellfish Research,* 23, 51-61.

[6] Anderson, D.M., Alpermann, T.J., Cembella, A.D., Collos, Y., Masseret, E. and Montresor, M. 2012. The globally distributed genus Alexandrium: Multifaceted roles in marine ecosystems and impacts on human health. *Harmful Algae,* 14, 10–35.

[7] James, K.J., Carey, B., O'Halloran, J.,Van Pelt, F.N.A.M. and Skrabakova, Z. 2010. Shellfish toxicity: human health implications of marine algal toxins. *Epidemiol Infect,* 138, 927–940.

[8] Hallegraeff, G.M. 2010. Ocean Climate Change, Phytoplankton Community Responses, and Harmful Algal Blooms: A Formidable Predictive Challenge. *J. Phycol.,* 46, 220–235.

[9] Picot, C., Nguyen, T. A., Roudot, A. C. and Parent-Massin, D. 2011. A Preliminary Risk Assessment of Human Exposure to Phycotoxins in Shellfish: A Review, Human and Ecological Risk Assessment. *An International Journal,* 17, 328-366.

[10] Gerssen, A., Pol-Hofstad, I., Poelman. M., Mulder, P., van den Top, H. and de Boer, J. 2010. Marine Toxins: Chemistry, Toxicity, Occurrence and Detection, with Special Reference to the Dutch Situation. *Toxicon,* 2, 878-904.

[11] Nakamura, Y. and Kerciku F. 2000. Effects of filter-feeding bivalves on the distribution of water quality and nutrient cycling in a eutrophic coastal lagoon. *J. Mar. Syst.,* 26, 209-221.

[12] Morgan, K.L., Larkin, S.L. and Adams, C.M. 2009. Firm-level economic effects of HABS: A tool for business loss assessment. *Harmful Algae,* 8, 212–218.

[13] [EU] European Commission Regulation No 15/2011. 2011. Regulation (EC) No 2074/(2005) as regards recognised testing methods for detecting marine biotoxins in live bivalve molluscs. *Off J. Eur. Union,* L6/3–L6/6.

[14] Fernandez, M.L., Miguez, A., Cacho, E. and Martinez, A. 1996. Detection of okadaic acid esters in the hexane extracts of Spanish mussels. *Toxicon,* 34, 381–387.

[15] García, C., Truan, D., Lagos, M., Santelices, J.P., Díaz, J.C. and Lagos, N. 2005. Metabolic transformation of dinophysistoxin-3 into dinophysistoxin-1 causes human intoxication by consumption of O-acyl-derivatives dinophysistoxins contaminated shellfish. *J. Toxicol. Sci.,* 30, 287–296.

[16] McCarron, P., Kilcoyne, J. and Hess, P. 2008. Effects of cooking and heat treatment on concentration and tissue distribution of okadaic acid and dinophysistoxin-2 in mussels (*Mytilus edulis*). *Toxicon,* 51,1081–1089.

[17] Bricelj, V.M., Connell, L., Konoki, K., MacQuarrie, S.P., Scheuer, T. and Catterall, W.A. 2005. Sodium channel mutation leading to saxitoxin resistance in clams increases risk of PSP. *Nature,* 434, 763–767.

[18] Llewellyn, L.E. 2006. Saxitoxin, a toxic marine natural product that targets a multitude of receptors. *Nat. Prod. Rep.,* 23, 200–222.

[19] Wiese, M., D'Agostino, P.M., Mihali, T.K., Moffitt, M.C. and Neilan, B.A. 2010. Neurotoxic Alkaloids: Saxitoxin and Its Analogs. *Mar. Drugs,* 8, 2185-2211.

[20] Perez, S., Vale, C., Botana, A.M., Alonso, E., Vieytes, M.R. and Botana, L.M. 2011. Determination of toxicity equivalent factors for paralytic shellfish toxins by electrophysiological measurements in cultured neurons. *Chem. Res. Toxicol.,* 24, 1153-1157.

[21] Donovan, C.J., Ku, J.C., Quilliam, M.A. and Gill, T.A. 2008. Bacterial degradation of paralytic shellfish toxins. *Toxicon,* 52, 91–100.

[22] Kodama, M. 2010. Paralytic Shellfish Poisoning Toxins: Biochemistry and Origin. *Aqua-BioScience Monographs,* 3, 1–38.

[23] van de Riet, J., Gibbs, R.S., Muggah, P.M., Rourke, W.A. Macneil, J.D. and Quilliam, M.A. 2011. Liquid Chromatography Post-Column Oxidation (PCOX) Method for the Determination of Paralytic Shellfish Toxins in Mussels, Clams, Oysters, and Scallops: Collaborative Study. *J. AOAC Int.,* 94, 1154–1176.

[24] Gessner, B.D., Bell, P., Doucette, G.J., Moczydlowski, E., Poli, M.A., Van Dolah, F. and Hall, S. 1997. Hypertension and identification of toxin in human urine and serum following a cluster of mussel-associated paralytic shellfish poisoning outbreaks. *Toxicon,* 35, 711–722.

[25] García, C., del Carmen Bravo, M., Lagos, M. and Lagos, N. 2004. Paralytic shellfish poisoning: Post-mortem analysis of tissue and body fluid samples from human victims in the Patagonia fjords. *Toxicon,* 43, 149–158.

[26] Etheridge, S.M. 2010. Paralytic shellfish poisoning: Seafood safety and human health perspectivas. *Toxicon,* 56, 108–122.

[27] Deeds, J., Landsberg, J., Etheridge, S., Pitcher, G. and Longan, S. 2008. Non-traditional vectors for paralytic shellfish poisoning. *Mar. Drugs,* 6, 308–348.

[28] Anderson, D.M., Cembella, A.D. and Hallegraeff, G.M. 2012. Progress in understanding harmful algal blooms: paradigm shifts and new technologies for research, monitoring, and management. *Ann. Rev. Mar. Sci.,* 4, 143-76.

[29] Montoya, N.G., Fulco, V.K., Carignan, M.O. and Carreto, J.I. 2010. Toxin variability in cultured and natural populations of *Alexandrium tamarense* from southern South America - Evidences of diversity and environmental regulation. *Toxicon*, 56, 1408-1418.

[30] Guzmán, L., Pacheco, H., Pizarro, G., Alarcón, C. 2002. *Alexandrium catenella* y veneno paralizante de los mariscos en Chile. In: Sar, E.A., Ferrario, M.E. and Reguera, B. (Eds.), Floraciones Algales Nocivas en el Cono Sur Americano. Instituto Español de Oceanografía (pp. 235-256). Madrid, España.

[31] Lagos, N. 2003. Paralytic shellfish poisoning phycotoxins: occurrence in South America. *Comments Toxicol.,* 9, 175–193.

[32] Menezes, M., Branco, S., Proença, L.A.O. and Schramm, M.A. 2007. Bloom of *Alexandrium minutum* Halim on Rio de Janeiro coast: occurrence and toxicity. Harmful Algae News 34, 7–9.

[33] Etheridge, SM. and Roesler, C.S. 2005. Effects of temperature, irradiance, and salinity on photosynthesis, growth rates, total toxicity, and toxin composition for *Alexandrium fundyense* isolates from the Gulf of Maine and Bay of Fundy. *Deep-Sea Research II. 52*, 2491–2500.

[34] Compagnon, D., Lembeye, G., Marcos, N., Ruiz-Tagle, N., Lagos, N. 1998. Accumulation of paralytic shellfish poisoning toxins in the bivalve *Aulacomya ater* and two carnivorous gastropods *Concholepas concholepas* and *Argobuccinum ranelliformes* during an *Alexandrium catenella* bloom in southern Chile. *J. Shellfish Res.,* 17, 67–73.

[35] Krock, B., Seguel, C.G. and Cembella, A.D. 2007. Toxin profile of *Alexandrium catenella* from the Chilean coast as determined by liquid chromatography with fluorescence detection and liquid chromatography coupled with tandem mass spectrometry. *Harmful Algae,* 6, 734–744.

[36] Oshima, Y., Bolch, C.J. and Hallegraeff, G.M. 1992. Toxin composition of resisting cysts of *Alexandrium tamarense* (Dinophyceae). *Toxicon,* 30, 1539–1544.

[37] Vale, P., Botelho, M.J., Rodrigues, S.M., Gomes, S.S. and Sampayo, M.A.D. 2008. Two decades of marine biotoxin monitoring in bivalves from Portugal (1986–2006): A review of exposure assessment. *Harmful Algae,* 7, 11–25.

[38] Molinet, C., Lafon, A., Lembeye, G. and Moreno, C. 2003. Patrones de distribución espacial y temporal de floraciones de *Alexandrium catenella* (Whoedon and Kofoid)

Balech 1985, en aguas interiores de la Patagonia noroccidental de Chile. *Revista Chilena de Historia Natural,* 76, 681–698.

[39] Jellet, J.F. 1993. Phycotoxins and shellfish aquaculture. *World aquaculture,* 24, 32-44.

[40] Vila, M., Garces, E., Maso, M. and Camp, J. 2001. Is the distribution of the toxic dinoflagellate *Alexandrium catenella* expanding along the NW Mediterranean Coast? *Mar. Ecol. Prog. Ser.,* 222, 73–83.

[41] Eilertsen, H. and Wyatt, T. 1988. A model of Alexandrium population dynamics. In: Reguera B., Blanco, J., Fernandez, M.I. and Wyatt, T. (eds). Harmful Algae: (pp. 196-199). Xunta de Galicia and Intergovernmental Oceanographic Commission of UNESCO. Vigo, Spain.

[42] Tsujino, M., Kamiyama, T., Uchida, T., Yamaguchi, M. and Itakura, S. 2002. Abundance and germination capability of resting cyst of *Alexandrium* spp. (Dinophyceae) from faecal pellets of macrobenthic organism. *Journal of Experimental Marine Biology and Ecology,* 271, 1-7.

[43] Itakura S. and Yamahuchi, M. 2001. Germination characteristics of naturally ocurring cyts of *Alexandium tamarense* (Dinophyceae) in Hirsoshima bay, *Inland Sea of Japan. Phycologia,* 40, 263-267.

[44] Guéguen, M., Baron, R., Bardouil, M., Truquet, P., Haberkorn, H., Lassus, P., Barillé, L. and Amzil, Z. 2011. Modelling of paralytic shellfish toxin biotransformations in the course of *Crassostrea gigas* detoxification kinetics. *Ecological Modelling,* 222, 3394–3402.

[45] Kwong, R.W., Wang, W.X., Lam, P.K. and Yu, P.K. 2006. The uptake, distribution and elimination of paralytic shellfish toxins in mussels and fish exposed to toxic dinoflagellates. *Aquat. Toxicol.,* 80, 82-91.

[46] Jaime, E., Gerdts, G. and Luckas, B. 2007. In vitro transformation of PSP toxins by different shellfish tissues. *Harmful Algae,* 6, 308–316.

[47] Gainey, L.F. and Shumway, S.E. 1988. A compendium of the responses of bivalve mollusks to toxic dinoflagellates. *Journal of Shellfish Research,* 7, 626–628.

[48] Connell, L.B., Macquarie, S.P., Twarog, B.M., Iszard, M. and Bricelj M. 2007. Population diferences in nerve resistance to paralytic shellfish toxins in softshell clam, *Mya arenaria*, associated with sodium channel mutations. *Mar. Biol.,* 150, 1227–1236.

[49] Hégaret, H., Wikfors, G.H., Soudant, P., Lambert, C., Shumway, S.E., Bérard, J.B. and Lassus, P. 2007. Toxic dinoflagellates (*Alexandrium fundyense* and *A. catenella*) have minimal apparent effect on oyster hemocytes. *Marine Biology,* 152, 441–447.

[50] May, S.P., Burkholder, J.A.M., Shumway, S.E., Wikfors, G.H., Frank, D., Hsia, M., Hégaret, H. and Dorch, Q. 2010. Effects of the toxic dinoflagellate *Alexandrium monilatum* on survival, grazing and behavioral response of three ecologically important shellfish species. *Harmful Algae,* 9, 281–293.

[51] Maki, H. 2002. Origins of spontaneous mutations: specicity and directionality of base-substitution, frameshift and sequence-substitution mutagenesis. *Annu. Rev. Genet.,* 36, 279–303.

[52] Twarog, B.M., Hidaka, T. and Yamaguchi, H. 1972. Resistance to tetrodotoxin and saxitoxin in nerves of bivalve molluscs: a possible correlation with paralytic shellfish poisoning. *Toxicon,* 10, 273–278.

[53] Shumway, S.E. and Gainey, L. 1992. A review of the physiological effects of toxic dinoflagellates on bivalve molluscs. Proc. Ninth Int. Malac. Congress: 357-362.

[54] Bricelj, V.M. and Shumway, S.E. 1998. Paralytic shellfish toxins in bivalve molluscs: occurrence, transfer kinetics and biotransformation. *Fisheries Science,* 6, 315–383.

[55] Smith, E.A., Grant, F., Ferguson, C.M.J. and Gallacher, S. 2001. Biotransformations of paralytic shellfish toxins by bacteria isolated from bivalve molluscs. *Applied and Environmental Microbiology,* 67, 2345–2353.

[56] Oshima, Y. 1995. Postcolumn derivatization liquid chromatographic method for paralytic shellfish toxins. *J. AOAC Int.,* 78, 528–532.

[57] Bricelj, V.M., Lee, J.H., Cembella, A.D. and Anderson, D.M. 1990. Uptake kinetics of paralytic shellfish toxins from the dinoflagellate *Alexandrium fundyense* in the mussel *Mytilus edulis. Mar. Ecol. Prog. Ser.,* 63, 177–188.

[58] Shumway, S. 1995. Phycotoxin related shellfish poisoning: Bivalve molluscs are not the only vectors. *Reviews in Fisheries Science,* 3, 1-31.

[59] Li, A.MY., Yu, P.K.N., Hsieh, D.P.H., Wang, W.-X., Wu, R.S.S. and Lam, P.K.S. 2005. Uptake and depuration paralytic shellfish toxins in the green-lipped mussles, *Perna viridis*: a dynamic model. *Environ. Toxicol. Chem.,* 24, 129-135.

[60] Paz, B., Daranas, A.H., Norte, M., Riobo, P., Franco, J.M. and Fernández, J.J. 2008. Yessotoxins, a group of marine polyether toxins: an overview. *Mar. Drugs,* 6, 73–102.

[61] Dominguez, H.J., Paz, B., Daranas, A.H., Norte, M., Franco, J.M. and Fernández, J.J. 2010. Dinoflagellate polyether within the yessotoxin, pectenotoxin and okadaic acid toxin groups: characterization, analysis and human health implications. *Toxicon,* 56, 191–217.

[62] Glibert, P.M., Burkholder, J.M and Kana, T.M. 2012. Recent insights about relationships between nutrient availability, forms, and stoichiometry, and the distribution, ecophysiology, and food web effects of pelagic and benthic Prorocentrum species. *Harmful Algae,* 14, 231–259.

[63] Reguera, B., Velo-Suárez, L., Raine, R. and Park, M.G. 2012. Harmful Dinophysis species: A review. *Harmful Algae,* 14, 87–106.

[64] Kat, M. 1979. The occurrence of Prorocentrum species and coincidental gastrointestinal illness of mussel consumers. In Taylor, D.L. and Seliger, H.H. (Eds), Toxic Dinoflagellate Blooms (pp. 215-220) New York, , USA: Elsevier.

[65] García, C., Gonzalez, V., Cornejo, C., Palma-Fleming, H. and Lagos, N. 2004. First evidence of dinophysistoxin-1 ester and carcinogenic polycyclic aromatic hydrocarbons in smoked bivalves collected in the Patagonia fjords. *Toxicon,* 43, 121–131.

[66] Elgarch, A., Vale, P., Rifai, S. and Fassouane, A. 2008. Detection of diarrheic shellfish poisoning and azaspiracid toxins in Moroccan mussels: Comparison of the LC-MS method with the commercial immunoassay kit. *Mar. Drugs,* 6, 587–594.

[67] García, C., Rodriguez-Unda, N., Contreras, C., Barriga, A. and Lagos, N. 2012. Lipophilic toxin profiles detected in farmed and benthic mussels populations from the most relevant production zones in Southern Chile. *Food Addit Contam Part A.,* 29, 1011-1020.

[68] Yasumoto, T., Murata, M. and Oshima, Y. 1985. Diarrhetic shellfish toxins. *Tetrahedron,* 41, 1019–1025.

[69] Fux, E., Smith, J.L., Tong, M., Guzmán, L. and Anderson, D.M. 2011. Toxin profiles of five geographical isolated of *Dinophysis* spp. From North and South America. *Toxicon,* 57, 275–287.

[70] Hamano, Y., Kinoshita, Y. and Yasumoto, T. 1986. Enteropathogenicity of diarrhetic shellfish toxins in intestinal models. *J. Food Hyg. Soc. Jpn.,* 27, 375-379.

[71] Suganuma, M., Fujiki, H., Suguri, H., Yoshizawa, S., Hirota, M., Nakayasu, M., Ojika, M., Wakamatzu, K., Yamada, K. and Sugimura T. 1988. Okadaic acid: an additional non-phorbol-12-tetradecanoate-13-acetate-type tumor promoter. *Proc. Natl. Acad. Sci.,* 85, 1768–1771.

[72] Fujiki, H., Suganuma, M., Suguri, H., Yoshizawa, S., Takagi, K., Uda, N., Wakamatsu, K., Yamada, K. and Murata, M. 1988. Diarrhetic shellfish toxin, dinophysistoxin-1, is a tumor promoter on mouse skin. *Jpn. J. Cancer Res.,* 79, 1089–1093.

[73] Cordier, S., Monfort, C., Miossec, L., Richardson, S. and Belin C. 2000. Ecological analysis of digestive cancer mortality related to contamination by diarrhetic shellfish poisoning toxins along the coasts of France. *Environ Res.,* 84, 145–150.

[74] Maneiro, E., Victorio, R., Costas, E. and Hernandez, J.M. 2008. Shellfish consumption: a major risk factor for colorectal cancer. *Med. Hypothesis,* 70, 409–412.

[75] Huhn, J., Jeffrey, P.D., Larsen, K., Rundberget, T., Rise, F., Cox, N.R., Arcus, V., Shi, Y. and Miles, C.O. 2009. A structural basis for the reduced toxicity of dinophysistoxin-2. *Chem. Res. Toxicol,* 22, 1782-1786.

[76] Hayat, A., Barthelmebs, L. and Marty J.L. 2012. A Simple Colorimetric Enzymatic-Assay for Okadaic Acid Detection Based on the Immobilization of Protein Phosphatase 2A in Sol-Gel. *Appl. Biochem. Biotechnol.,* 166, 47-56.

[77] Van den Heuvel, J.P. 2002. Xenobiotic receptor systems: Introduction and overview. In: Van den Heuvel, JP., Perdew, G.H., Mattes, W.B. and Greenlee, W.F. (Eds), Comprehensive Toxicology (pp. 646). Amsterdam, Holland: Elsevier Science BV.

[78] Vale, P. 2006. Detailed profiles of 7-O-acyl esters in plankton and shellfish from the Portuguese coast. *Journal of Chromatography A,* 1128, 181–188.

[79] Landsberg, J.H. 2002. The effects of harmful algal blooms on aquatic organisms. *Rev. Fish. Sci.,* 10, 113–390.

[80] Torgersen, T.; Miles, C.; Rundberget, T. and Wilkins, A. 2008. New esters of okadaic acid in seawater and blue mussels (*Mytilus edulis*). *Agric. Food Chem.,* 56, 9628–9635.

[81] Torgersen, T., Sandvik, M., Lundve, B. and Lindegarth, S. 2008. Profiles and levels of fatty acid esters of okadaic acid group toxins and pectenotoxins during toxin depuration. Part II: blue mussels (*Mytilus edulis*) and flat oyster (*Ostrea edulis*). *Toxicon,* 52, 418–427.

[82] Manfrin, C., Dreos, R., Battistella, S., Beran, A., Gerdol, M., Varotto, L., Lanfranchi, G., Venier, P. and Pallavicini A. 2010. Mediterranean mussel gene expression profile induced by okadaic acid exposure. *Environ. Sci. Technol.,* 44, 8276-8283.

[83] Svensson, S. and Förlin, L. 2004. Analysis of the importance of lipid breakdown for elimination of okadaic acid (diarrhetic shellfish toxin) in mussels, *Mytilus edulis*: results from a field study and a laboratory experiment. *Aquat. Toxicol.,* 66, 405-418.

[84] Reboreda, A., Lago, J., Chapela, M.J., Vieites, J.M., Botana, L.M., Alfonso, A. and Cabado, A.G. 2010. Decrease of marine toxin content in bivalves by industrial processes. *Toxicon,* 55, 235–243.

[85] EFSA, 2008. Opinion of the scientific panel on contaminants in the food chain on a request from the European Commission on marine biotoxins in shellfish-okadaic acid and analogues. *The EFSA Journal,* 589, 1–62.

[86] European Union Parliament. 2009. List of member states`autohorizations of food and food ingredients which may be treated with ionization radiation. *Off. J. Eur. Union.,* C 283; 52, 5.

[87] Raúl, F., Gosse, F., Delince, H., Hartwing, A., Marchioni, E., Miesch, M., Werner, D. and Burnouf, D. 2002. Food-borne radiolytic compounds (2-alkylcyclobutanones) may promote experimental colon carcinogenesis. *Nutr. Cancer,* 44, 189-191.

[88] Jørgensen, K., Scanlon, S. and Jensen, L.B. 2005. Diarrhetic shellfish poisoning toxin esters in Danish blue mussels and surf clams. *Food Addit. Contam.,* 22, 743–751.

[89] Svensson, S., Särngren, A. and Förlin, L. 2003. Mussel blood cells, resistant to the cytotoxic effects of okadaic acid, do not express cell membrane p-glycoprotein activity (multixenobiotic resistance). *Aquatic Toxicology,* 65, 27-37

[90] García, C., Schonstedt, V., Santelices, J.P. and Lagos, N. 2006. High amount of dinophysistoxin-3 in *Mytilus chilensis* collected in Seno de Reloncaví, Chile, during massive human intoxication associated with outbreak of Vibrio parahaemolyticus. *J. Toxicol. Sci.,* 31, 305–314.

[91] Chambers, T.C., Raynor, R.L. and Kuo, J.F. 1993. Multidrug-resistant human KB carcinoma cells are highly resistant to the protein phosphatase inhibitors okadaic acid and calyculin A. Analysis of potential mechanisms involved in toxin resistance. *Int. J. Cancer,* 53, 323-327.

[92] Cornwall, R., Holland Toomey, B., Bard, S., Bacon, C., Jarman, W.M. and Epel, D. 1995. Characterization of multi-xenobiotic/multidrug transport in the gills of the mussel *Mytilus californianus* and identification of environmental substrates. Aquat. Toxicol. 31, 277-296.

[93] Kurelec, B. 1995. Reversion of the multixenobiotic resistance mechanism in gills of a marine mussel *Mytilus galloprovincialis* by a model inhibitor and environmental modulators of P170-glycoprotein. *Aquat. Toxicol.,* 33, 93-103.

[94] Hu, T., Curtis, J.M., Walter, J.A., McLachlan, J.L. and Wright, J.L.C. 1995. Two new water-soluble DSP toxin derivatives from the dinoflagellate *Prorocentrum maculosum*: possible storage and excretion products. *Tetrahedron Lett.,* 36, 9273–9276.

[95] Hu, T., Curtis, J.M., Walter, J.A. and Wright, J. 1995. Identification of DTX-4, a new water-soluble phosphatase inhibitor from the toxic dinoflagellate *Prorocentrum lima. J. Chem. Soc. Chem. Comm.,* 5, 597–599.

[96] Eiki, K., Satake, M., Koike, K., Ogata, T., Mitsuya, T. and Oshima, Y. 2005. Confirmation of yessotoxin production by the dinoflagellate *Protoceratium reticulatum* in Mutsu Bay. *Fish Sci.,* 71, 633-638.

[97] Ciminiello, P., Dell'Aversano, C., Fattorusso, E., Forino, M., Magno, S., Guerrini, F., Pistocchi, R. and Boni, L. 2003. Complex yessotoxins profile in *Protoceratium reticulatum* from north-western Adriatic sea revealed by LC–MS analysis. *Toxicon,* 42, 7–14.

[98] Paz, B., Daranas, A.H., Cruz, P.G., Franco, J.M., Pizarro, G., Souto, M.L., Norte, M. and Fernandez, J.J. 2007. Characterisation of okadaic acid related toxins by liquid chromatography coupled with mass spectrometry. *Toxicon,* 50, 225–235.

[99] Suzuki, T., Satake, M., Yoshimatsu, S., Oshima, Y., Horie, Y., Koike, K. and Iwataki, M. 2007. Yessotoxin analogues in several strains of *Protoceratium reticulatum* in Japan determined by liquid chromatography-hybrid triple quadrupole/linear ion trap mass spectrometry. *J. Chromatogr. A,* 1142, 172-177.

[100] Vershinin, A., Moruchkov, A., Morton, S.L., Leighfield, T.A., Quilliam, M.A., Ramsdell, J.S. 2006. Phytoplankton composition of the Kandalaksha Gulf, Russian White Sea: Dinophysis and lipophilic toxins in the blue mussel (*Mytilus edulis*). *Harmful Algae,* 5, 558-564.

[101] Samdal, I.A., Naustvoll, L.J., Olseng, C.D., Briggs, L.R. and Miles, C.O. 2004. Use of ELISA to identify *Protoceratium reticulatum* as a source of yessotoxin in Norway. *Toxicon,* 44, 75-82.

[102] Yasumoto, T., Takizawa, A. 1997. Fluorometric measurement of yessotoxins in shellfish by high pressure liquid chromatography. *Biosci. Biotechnol. Biochem.,* 61, 1775-1777.

[103] Rossini, G.P. and Hess, P. 2010. Phycotoxins: chemistry, mechanisms of action and shellfish poisoning. EXS, 100, 65–122.

[104] Miles, C. O., Samdal, I.A., Aasen, J.A.G., Jensen, D.J., Quilliam, M.A., Petersen, D., Briggs, L.R., Wilkins, A.L., Rise, F., Cooney, J.M. and MacKenzie, L. 2005. Evidence for numerous analogs of yessotoxin in *Protoceratium reticulatum*. *Harmful Algae*, 4, 1075-1091.

[105] Rhodes, L., McNabb, P., de Salas, M., Briggs, L., Beuzenberg, V. and Gladstone M. 2006. Yessotoxin production by *Gonyaulax spinifera*. *Harmful Algae,* 5, 148–155.

[106] Ogino, H., Kumagai, M. and Yasumoto, T. 1997. Toxicologic evaluation of yessotoxin. *Nat. Toxins,* 5, 255-259.

[107] Tubaro, A., Sosa, S., Carbonatto, M., Altinier, G., Vita, F., Melato, M., Satake, M. and Yasumoto, T. 2003. Oral and intraperitoneal acute toxicity studies of yessotoxin and homoyessotoxins in mice. *Toxicon,* 41, 783-792.

[108] Alfonso, A., de la Rosa, L.A., Vieytes, M.R., Yasumoto, T. and Botana, L.M. 2003. Yessotoxin, a novel phycotoxin, activates phosphodiesterase activity. Effect of yessotoxin on cAMP levels in human lymphocytes. *Biochem. Pharmacol.,* 65, 193-208.

[109] Aune, T., Sorby, R., Yasumoto, T., Ramstad, H. and Landsverk, T. 2002. Comparison of oral and intraperitoneal toxicity of yessotoxin towards mice. *Toxicon,* 40, 77–82.

[110] Alexander, J., Audunsson, G.A., Benford, D., Cockburn, A., Cradevi, J.P., Dogliotti, E., Domenico, A.D., Fernandez-Cruz, M.L., Fink-Gremmels, J., Furst, P., Galli, C., Grandjean, P., Gzyl, J., Heinemeyer, G., Johansson, N., Mutti, A., Schlatter, J., van Leeuwen, R., van Peteghem, C. and Verger, P. 2008. Marine biotoxins in shellfish - okadaic acid and analogues. *EFSA J.,* 589, 1–62.

[111] Miles, C.O. 2006. Identification of fatty acid esters of pectenotoxin-2 seco acid in Blue mussels (*Mytilus edulis*) from Ireland. *J. Agric. Food Chem.,* 54, 5672–5678.

[112] Aasen, J., Samdal, I. A., Miles, C. O., Dahl, E., Briggs, L. R. and Aune, T. 2005. Yessotoxins in Norwegian blue mussels (*Mytilus edulis*): uptake from *Protoceratium reticulatum*, metabolism and depuration. *Toxicon,* 45, 265-272.

[113] Suzuki, T. and Quilliam, M.A. 2011. LC-MS/MS analysis of diarrhetic shellfish poisoning (DSP) toxins, okadaic acid and dinophysistoxin analogues, and other lipophilic toxins. *Anal. Sci.,* 27, 571– 584.

[114] Ciminiello, P., Fattorusso, E., Forino, M., Magno, S., Poletti, R. and Viviani, R. 1999. Isolation of 45-hydroxyyessotoxin from mussels of the Adriatic Sea. *Toxicon,* 37, 689–693.

[115] Ciminiello, P., Fattorusso, E., Forino, M., Poletti, R. and Viviani, R. 2000. Structure determination of carboxyhomoyessotoxin, a new yessotoxin analog isolated from Adriatic mussels. *Chem. Res. Toxicol.,* 13, 770–774.

[116] Ciminiello, P., Fattorusso, E., Forino, M. and Poletti, R. 2001. 42,43,44,45,46,47, 55-Heptanor-41-oxohomoyessotoxin, a new biotoxin from mussels of the Northern Adriatic Sea. *Chem. Res. Toxicol.,* 14, 596–599.

[117] Li, Z., Mengmeng, G., Shouguo, Y., Qingyin, W. and Zhijun, T. 2010. Investigation of pectenotoxins profiles in the Yellow Sea (China) using a passive sampling technique. *Mar. Drugs,* 8, 1263– 1272.

[118] Miles, C.O., Wilkins, A.L., Munday, R., Dines, M.H., Hawkes, A.D., Briggs, L.R., Sandvik, M., Jensen, D.J., Cooney, J.M., Holland, P.T., Quilliam, M.A., MacKenzie, A.L. and Beuzenberg, V. 2004. Isolation of pectenotoxin-2 from *Dinophysis acuta* and its conversion to pectenotoxin-2 seco acid, and preliminary assessment of their acute toxicities. *Toxicon,* 43, 1–9.

[119] Yasumoto, T., Murata, M., Oshima, Y., Sano, M., Matsumoto, G. and Clardy, J. 1985. Diarrhetic shellfish toxins. *Tetrahedron,* 41, 1019–1025.

[120] Krock, B., Seguel, C.G., Valderrama, K. and Tillmann, U. 2009. Pectenotoxins and yessotoxin from Arica Bay, North Chile as determined by tandem mass spectrometry. *Toxicon,* 54, 364-367.

[121] Vale, P. and Sampayo, M.A. 2002. Pectenotoxin-2 seco acid, 7-epi-pectenotoxin-2 seco acid and pectenotoxin-2 in shellfish and plankton from Portugal. *Toxicon,* 40, 979–987.

[122] Aasen, J.A., Hardstaff, W., Aune, T. and Quilliam, M.A. 2006. Discovery of fatty acid ester metabolites of spirolide toxins in mussels from Norway using liquid chromatography/tandem mass spectrometry. Rapid Commun. *Mass Spectrom.,* 20, 1531–1537.

[123] Suzuki, T., Igarashi, T., Ichimi, K., Watai, M., Suzuki, M., Ogiso, E. and Yasumoto, T. 2005. Kinetics of diarrhetic shellfish poisoning toxins, okadaic acid, dinophysistoxin-1, pectenotoxin-6 and yessotoxin in scallops *Patinopecten yessoensis*. *Fish. Sci.,* 71, 948–955.

[124] Miles, C.O. 2007. Pectenotoxins. In: Botana, L. (Ed.), Phycotoxins: Chemistry and Biochemistry (pp. 159-186). Blackwell Publishing.

[125] Burguess, V. and Shaw, G. 2001. Pectenotoxins – an issue for public health. A review of their comparative toxicology and metabolism. *Environment International,* 27, 275-283.

[126] Jung, I.H., Sim, C.J. and Lee, C.O. 1995. Cytotoxic compounds from a two-sponge association. *J. Nat. Prod.,* 58, 1722-1726.

[127] Suzuki, T., Mackenzie, L., Stirling, D. and Adamson, J. 2001. Pectenotoxin-2 seco acid: a toxin converted from pectenotoxin-2 by the New Zealand greenshell mussel *Perna canaliculus*. *Toxicon,* 39, 507-514.

[128] James, K. J., Sierra, M. D., Lehane, M., Braña Magdalena, A., Moroney, C. and Furey, A. 2004. Azaspiracid poisoning: Aetiology, toxin dynamics and bioconversion in shellfish. In: Harmful Algae 2002, Steidinger, K., Landsberg, J., Tomas, C. R. and Vargo, G. A., (Eds). Florida Fish and Wildlife Conservation Commission, Florida Institute of Oceanography and Intergovernmental Oceanographic Commission of UNESCO (pp 111-112). Florida, USA.

[129] Tillman, U., Elbrachter, M., Krock, B., John, U. and Cembella, A. 2009. *Azadinium spinosum* gen. et sp. nov. (Dinophycae) identified as primary producer of azaspiracid toxins. *European Journal of Phycology,* 44, 63–79.

[130] Krock, B., Tillmann, U., John, U. and Cembella, A.D. 2009. Characterization of azaspiracids in plankton size-fractions and isolation of an azaspiracid- producing dinoflagellate from the North Sea. *Harmful Algae,* 8, 254–263.

[131] Rehmann, N., Hess, P. and Quilliam, M.A. 2008. Discovery of new analogs of the marine biotoxin azaspiracid in blue mussels *Mytilus edulis* by ultra-performance liquid chromatography/tandem mass spectrometry. *Rapid Communications in Mass Spectrometry,* 22, 549–558.

[132] Satake, M., Ofuji, K., Naoki, H., James, K.J., Furey, A., McMahon, T., Silke, J. and Yasumoto, T. 1998. Azaspiracid, a new marine toxin having unique spiroring assemblies, isolated from Irish mussels, *Mytilus edulis. Journal of the American Chemical Society,* 120, 9967–9968.

[133] Ofuji, K., Satake, M., Oshima, Y., McMahon, T., James, K. J. and Yasumoto, T. 1999. A sensitive and specific determination method for azaspiracids by liquid chromatography mass spectrometry. *Nat. Toxins,* 7, 247-250.

[134] Amzil, Z., Sibat, M., Royer, F. and Savar, V. 2008. First report of azaspiracid and yessotoxin groups detection in French shellfish. *Toxicon,* 52, 39–48.

[135] Twiner, M.J., Rehmann, N., Hess, P. and Doucette, G.J. 2008. Azaspiracid shellfish poisoning: a review on the chemistry, ecology and toxicology with an emphasis on human health impact. *Mar. Drugs,* 6, 39–72.

[136] Lopez-Rivera, A., O'Callaghan, K., Moriarty, M., O'Driscoll, D., Hamilton, B., Lehane, M., James, K.J. and Furey, A. 2010. First evidence of azaspiracids (AZAs): A family of lipophilic polyether marine toxins in scallops (*Argopecten purpuratus*) and mussels (*Mytilus chilensis*) collected in two regions of Chile. *Toxicon,* 55, 692–701.

[137] James, K.J., O'Driscoll, D., Fernandez, J.G., Furey, A. 2008. Azaspiracids: chemistry, bioconversion and determination. In: Botana, L.M. (Ed.), Seafood and Freshwater Toxins: Pharmacology, Physiology and Detection (pp. 763-773). New York, USA: CRC Press, Taylor and Francis.

[138] McCarron, P., Giddings, S.D., Miles, C.O. and Quilliam, M.A. 2011. Derivatization of azaspiracid biotoxins for analysis by liquid chromatography with fluorescence detection. *Journal of Chromatography A,* 1218, 8089–8096.

[139] Twiner, M.J., Hess, P., Dechraoui, M.Y.B., McMahon, T., Samons, M.S., Satake, M., Yasumoto, T., Ramsdell, J.S. and Doucette, G.J. 2005. Cytotoxic and cytoskeletal effects of azaspiracid-1 on mammalian cell lines. *Toxicon,* 45, 891–900.

[140] Ronzitti, G., Hess, P., Rehmann, N. and Rossini, G.P. 2007. Azaspiracid-1 alters the E-cadherin pool in epithelial cells. *Toxicol. Sci,* 95, 427–435.

[141] James, K.J., Lehane, M., Moroney, C., Fernandez-Puente, P., Satake, M., Yasumoto, T. and Furey, A. 2002. Azaspiracid shellfish poisoning: unusual toxin dynamics in shellfish and the increased risk of acute human intoxications. *Food Addit. Contam.,* 19, 555-561.

[142] Ito, E. and Satake, M. 2002. Azaspiracid, a new marine toxin isolated from mussels: chemistry and histopathology. In: Fingerman, M. and Nagabhushanam, R. (Eds.), Recent Advances in Marine Biotechnology. Seafood Safety and Human Health, (vol. 7, pp. 29-39). New Hampshire, USA. Science Publishers.

[143] O'Driscoll, D., Skrabakov, Z., O'Halloran, J., van Pelt, F.N., James, K.J. 2011. Mussels Increase Xenobiotic (Azaspiracid) Toxicity Using a Unique Bioconversion Mechanism. *Environ. Sci. Technol.,* 45, 3102-3108

[144] Jauffrais, T., Marcaillou, C., Herrenknecht, C., Truquet, P., Séchet, V., Nicolau, E., Tillmann, U., Hess, P. 2012. Azaspiracid accumulation, detoxification and biotransformation in blue mussels (*Mytilus edulis*) experimentally fed *Azadinium spinosum*. *Toxicon,* 60, 582-595.

[145] Furey, A., O'Doherty, S. O'Callaghan, K. Lehane, M. and James, K.J. 2010. Azaspiracid poisoning (AZP) toxins in shellfish: Toxicological and health considerations. *Toxicon,* 56, 173–190,

[146] Ito, E., Satake, M., Ofuji, K., Kurita, N., James, K.J., McMahon, T. and Yasumoto, T. 2000. Multiple organ damage caused by a new toxin azaspiracid, isolated from mussels produced in Ireland. *Toxicon,* 38, 917–930.

[147] Anonymous. 2009. Marine biotoxins in shellfish–summary on regulated marine biotoxins. Scientific opinion on contaminants in the food chain. *The EFSA Journal,* 1–23.

[148] Nzoughet, J.K., Hamilton, J.T.G., Botting, C.H., Douglas, A., Devine, L., Nelson, J. and Elliot, C.T. 2009. Proteomics identification of azaspiracid toxin biomarkers in Blue Mussels, *Mytilus edulis*. *Molecular and Cellular Proteomics 8,* 1811–1822.

[149] Magdalena, A.B., Lehane, M., Moroney, C., Furey, A. and James, K.J. 2003. Food safety implications of the distribution of azaspiracids in the tissue compartments of scallops (*Pecten maximus*). *Food Addit. Contam.,* 20, 154–160.

[150] McCarron, P., Kilcoyne, J., Miles, C.O. and Hess, P. 2009. Formation of azaspiracids -3, -4, -6 and -9 via decarboxylation of carboxyazaspiracid metabolites from shellfish. *Journal of Agricultural and Food Chemistry,* 57, 160–169.

[151] Freites, L., Labarta, U. and Fernandez-Reiriz, M.J. 2002. Evolution of fatty acid profiles of subtidal and rocky shore mussel seed (*Mytilus galloprovincialis*, Link). Influence of envionmental parameters. *J. Exp. Mar. Biol. Ecol.,* 268, 185-204.

[152] Pazos, A.J., Sánchez, J.L., Román, G., Luz Pérez-Parallé, M. and Abad, M. 2003 Seasonal changes in lipid classes and fatty acid composition in the digestive gland of *Pecten maximus*. *Comp Biochem. Physiol. B Biochem. Mol. Biol.,* 134, 367-380.

[153] Rossignoli, A.E. and Blanco, J. 2010. Subcellular distribution of okadaic acid in the digestive gland of *Mytilus galloprovincialis*: first evidences of lipoprotein binding to okadaic acid. *Toxicon,* 55, 221–226.

[154] Suzuki, T. and Mitsuya, T. 2001. Comparison of dinophysistoxin-1 and esterified dinophysistoxin-1 (dinophysistoxin-3) contents in the scallop *Patinopecten yessoensis* and the mussel *Mytilus galloprovincialis*. *Toxicon,* 39, 905–908.

[155] Duinker, A., Bergslien, M., Strand, Ø., Olseng, C.D., Svardal, A. 2007. The effect of size and age on depuration rates of diarrhetic shellfish toxins (DST) in mussels (*Mytilus edulis* L.). *Harmful Algae,* 6, 288–300.

[156] Maestrini, S. 1998. Bloom dynamics and ecophysiology of *Dinophysis* spp. In: Anderson, D., Cembella, A. and Hallegraeff, G. (Eds.), Physiological Ecology of Harmful Algal Blooms (pp. 243-265). Ecological Sciences. Springer- Verlag.

[157] Sampayo, M.A., Alvito, P., Franca, S. and Sousa, I. 1990. *Dinophysis* spp. toxicity and relation to accompanying species. In: Granéli, E., Sundström B., Edler L. and Anderson D.M.M. (Eds.), Toxic Marine Phytoplankton. (pp. 215-220). New York, USA: Elsevier.

[158] Trefault, N., Krock, B., Delherbe, N., Cembella, A., Vásquez, M. 2011. Latitudinal transects in the southeastern Pacific Ocean reveal a diverse but patchy distribution of phycotoxins. *Toxicon,* 58, 389–397.

[159] García, C., Pruzzo, M., Rodriguez-Unda, N., Contreras, C., Lagos, N. 2010. First evidence of okadaic acid acyl-derivative and dinophysistoxin-3 in mussel samples collected in Chiloe Island, southern Chile. *J. Toxicol. Sci.,* 35, 335-344.

[160] Díaz, P., Molinet, C., Caceres, M.A., Valle-Levinson, A. 2011. Seasonal and intratidal distribution of *Dinophysis* spp. in a Chilean Fjord. *Harmful Algae,* 10, 155–164.

[161] Escalera, L., Pazos, Y., Doval, M.D., Reguera, B. 2012. A comparison of integrated and discrete depth sampling for monitoring toxic species of Dinophysis. *Mar. Pollut. Bull.,* 64, 106–113.

[162] Moroño, A., Arévalo, F., Fernández, M.L., Maneiro,J., Pazos, Y., Salgado, C., Blanco, J. 2003. Accumulation and transformation of DSP toxins in mussels *Mytilus galloprovincialis* during a toxic episode caused by *Dinophysis acuminata*. *Aquat. Toxicol.,* 62, 269–280.

In: Mussels: Ecology, Life Habits and Control
Editors: Jarek Nowak and Michal Kozlowski

ISBN: 978-1-62618-083-3

Chapter 3

MUSSELS AS A TOOL TO MONITOR POLLUTION

Elena Cubero-Leon and Corina M. Ciocan*

Environmental Biology Lecturer, School of Pharmacy and Biomolecular Sciences,
University of Brighton, Brighton, UK

ABSTRACT

In order to evaluate the impact and potential risk that toxicants have in the environment, biomonitoring programs are used. In general, they can fall into two categories, exposure and effects assessment. The traditional exposure assessment process involves the analytical measure of a target compound in the tissue of a sampled organism or in the water. Although exposure does not necessarily mean effects in the organisms, it can assist in confirming that the effects seen are due to anthropogenic impacts and not to natural alterations. Effects monitoring, on the other hand, looks at various levels of the biological organization to evaluate the status of the biological community. Mussels have been used as sentinel organisms to monitor effects and exposure to a variety of chemical contaminants in the aquatic environment such as metals, organochlorines, pesticides and PAHs and are being used in biomonitoring studies from international programmes like "Mussel Watch". Their sessile nature, wide geographical distribution, large populations, high filtering rates, and poor ability to metabolise xenobiotics make them ideal organisms for bioaccumulating contaminants from the aquatic environment and make them suitable as models in the field of environmental toxicology. This chapter focuses on the current status and perspectives on their usage as biological indicators to monitor pollution.

1. POLLUTION BIOMONITORING PROGRAMS

Life expectancy continues to increase with each decade, largely as a result of technological advances that have come from the synthesis of different chemicals such as fertilizers, pesticides and pharmaceuticals. The inappropriate handling of some of these

* Corresponding author: Dr. Corina Ciocan, Environmental Biology Lecturer, School of Pharmacy and Biomolecular Sciences, University of Brighton, UBH Priory HAP 207, Havelock Road, Hastings TN34 1BE , 01273 641185 Email: C.Ciocan@brighton.ac.uk.

chemicals however has cost significant degradation of the environment. Early evidence existed that several pesticides such as amitrole (Jukes & Shaffer, 1960) and dichlorodiphenyldichloroethane (DDT) (Burlington & Lindeman, 1950) interfered with the endocrine system of different organisms; but it was not until the publication of Rachel Carson's *Silent Spring* (1963) documenting detrimental effects of pesticides in the environment that widespread public concern was raised. Silent spring is considered to have played a crucial role in starting the modern environmental movement and led to an immediate strengthening of chemical regulations by regulatory agencies worldwide.

Since the mid 1970s, environmental monitoring became increasingly important mainly due to the awareness of environmental chemical pollution and human health risk. Nowadays, environmental risk assessments are widely used as a scientific tool to evaluate the probability and magnitude of biological effects resulting form exposure to chemical contaminants. The aim of risk assessments is to assist decision makers to tolerate, mitigate or eliminate the potential harm of the chemicals released to the environment. Risk assessment can be divided into risk analysis and risk management. Risk analysis generally includes three steps: exposure assessment, effects assessment and risk characterization. Risk management, more politically oriented, involves risk classification, risk-benefit analysis, risk reduction and risk monitoring (van der Oost el al., 2003). In the risk analysis stage, exposure assessment is the estimation of concentrations/doses to which populations are or may be exposed. Effects assessment on the other hand comprises hazard identification and dose-response assessment.

Traditional exposure assessment often relies on estimates of concentrations of various chemicals in a specific media (chemical monitoring) and the improvement of analytical technologies has made these measurements very sensitive. Nowadays very low levels of certain chemical can be detected in the environment. However organisms' internal dose can be much different due to specific bioavailability, absorption, disposition, metabolism and elimination. Lack of exposure data and a deficiency of understanding about important bioaccumulation mechanisms are major sources of scientific uncertainty in most risk assessments. Therefore chemical monitoring coupled with biomonitoring are considered useful tools in both exposure (through the measurement of contaminant levels in water and biota: bioaccumulation) and effects assessment (Figure 1). Biological monitoring or biomonitoring is the use of biological responses to assess changes in the environment, mainly due to anthropogenic causes. Biomonitoring data can provide not only a measure of an individual's internal dose to an environmental chemical (exposure assessment) but also information about the effects of the xenobiotic on the exposed organism building relationships between environmental pollutants and the occurrence of diseases (Smolders & Schoeters, 2007) (Figure 1).

Biomonitoring programs can be used to identify priority exposures of chemicals, recognize time trends in exposure, identify at-risk populations, establish reference ranges for comparison, provide integrated dose measurements and evaluate exposure intervention and prevention efforts (Sexton, 2004). These studies generally involve the measurements of the contaminants or biological changes (sometimes through the use of biomarkers) in indicator organisms, populations and/or communities. Suitable species for biomonitoring or indicator species (also called sentinel organisms) should spend the most sensitive phase of their life cycle in a single zone of pollution and they should be able to accumulate pollutants in their tissues from the environment or food. Biomonitoring programs can be qualitative, semi-quantitative or quantitative.

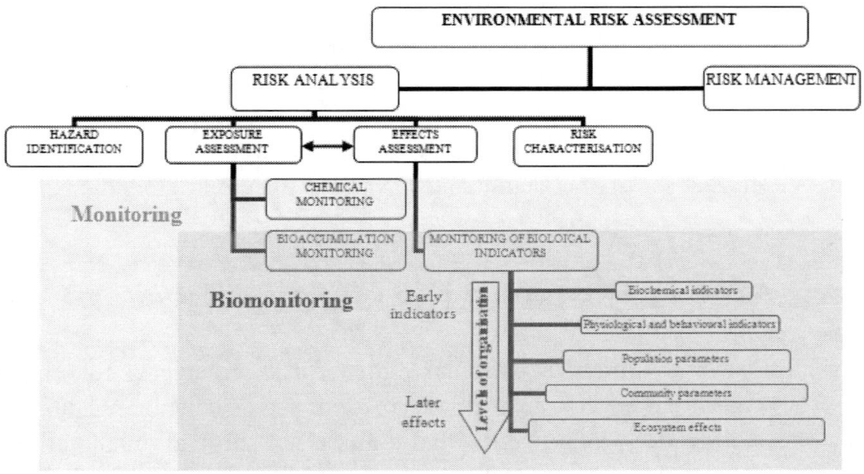

Based on van der Oost et al., (2003) and Landis & Yu (1995).

Figure 1. Relationship among the components of the risk assessment process and biomonitoring data.

In a qualitative study for example, the presence or absence of a specific species or communities may indicate pollution. However there are other factors that can produce the absence of species such as predation, geographical barriers or competition. Therefore, the absence of multiple species of different orders with similar tolerance levels that were present previously rather than absence of a single species is most indicative of pollution (Johnson et al., 1993). In quantitative and semi-quantitative studies a variety of measures can be used. The levels of a specific chemical compound or its metabolites can be measured in a certain tissue or body fluid such as urine or blood. Populations can also be assessed by means of statistical analysis: richness measures, abundances, diversity and evenness indices and/or biotic indices (Resh & Giap, 2010). Physiological or biochemical alterations can allow for early warning signs of disease as a consequence of exposure to environmental chemicals. They can be identified (qualitative assessment) and subsequently quantified by means of standard methodologies or criteria (quantitative or semi quantitative assessment).

In aquatic environments, a large number of studies have been performed using invertebrates, fish and microalgae as bio-indicators of pollution. Each group of organisms offers distinct abilities to indicate pollution in the long or short term but worldwide benthic macro invertebrates are the most widely used (Resh, 2007). The advantages of using benthic macro invertebrates such as mussels, clams, mayfly and crayfish is that they are easy to sample, they have limited mobility and they cannot escape toxic spills. They spend a large part of their life cycle in the same part of a stream and they are a critical part of the aquatic food web being indicators of cumulative effects of environmental variations.

Some examples of large scale biomonitoring programs are the Arctic Monitoring and Assessment Programme established in 1991 to measure the levels, and assess the effects of anthropogenic pollutants in all compartments of the Arctic Environment; the New York State's Biological Monitoring established in 1972 to evaluate the health of New York's rivers and streams through the analysis of macroinvertebrate communities; the Onchocerciasis Control Programme running for 29 years in 11 West African countries to biomonitor the effects of the insecticide used in 50,000 Km of African rivers between 1974 and 2002 that was used to control river blindness (onchocerciasis), a human disease transmitted by black flies; the International Mussel Watch Program (Goldberg, 1975), a program (still running) that was developed to detect biological contaminant trends in sediments and bivalve tissues with the objective to describe the current status of pollution and to detect changes in the environmental quality of estuarine and coastal waters around the world; the Niagara River mussel (*Elliptio complanata*) biomonitoring program between 1983-2009 conducted to determine the presence/absence of contaminants at various locations along the Niagara River.

2. Mussels as Indicator (Sentinel) Species

Mussels have several attributes that make them ideal organisms for environmental monitoring. They have wide geographical distributions and stable populations, they are sessile and suspension feeders, long-lived with a high rate of survival in laboratory and field studies, they are tolerant to environmental alterations and contaminants, they bioconcentrate pollutants and have very low levels of metabolizing enzyme activities.

Freshwater mussels (*Dreissena polymorpha*) and marine mussels (*Mytilus sp*) are the most common marine molluscs widely used as indicator organisms (Reeders et al., 1993). Because of their widespread distribution, their abundance and relatively well known biology, as well as their economic value in shellfish farming, mussels such as *Mytilus sp* have become bio-indicators of marine pollution. Mussels usually live in dense populations, with pure and hybrid individuals, in coastal areas of Europe, as well as North and South America. Mussels are recognised as being very tolerant to a wide variety of environmental conditions (salinity, oxygen, temperature, and desiccation), but they are not necessarily particularly tolerant of anthropogenic chemicals (Hawkins & Bayne, 1992; Holt et al., 1998). They have separate sexes, reproduction is external and larval stages are planktonic.

There are also good records of the locality and size of mussel beds, with some of the populations being mapped since 1950s or earlier (Seed & Suchanek, 1992). Mussel growth is affected by environmental conditions (including fluctuations in food quantity and quality, tidal exposure and chemical pollutants) and biological interactions such as parasitism (Venier et al, 2003). As a result, organismal performance (development, reproduction and mobility) may therefore be deeply modified. However most bivalve molluscs live alongshore which limits their use as indicator species of marine remote areas.

Biomonitoring programs in general can fall into two categories, exposure and effects (Landis & Yu 1995) (Figure 1). The traditional exposure biomonitoring process involves the analytical measure of a target compound in the tissue of a sampled organism. Effects biomonitoring, on the other hand, looks at various levels of the biological organization (from cellular to population levels) to evaluate the status of environmental stressors. Different methods have been described for the measurement of the xenobiotic effects at different levels of organization:

- Biochemical indicators: these methods include the measurements of stress proteins, DNA damage, DNA expression levels, metabolic indicators, metallothionein production and immunological suppression.
- Physiological and behavioural indicators: they include behavioural alterations, reproductive success, lesions and necrosis, tumours and compensatory behaviours.

- Population parameters: including population density, mating success and competitive alterations.
- Community parameters: measures diversity, competitive interactions and rates of change in community structures.
- Ecosystem effects: measures changes in structure, diversity and stability of ecosystems.

A variety of biochemical and physiological tests are being used nowadays as indicators and perhaps eventually as predictors of the effects of toxicants. The collective term that has been given to these measurements by the scientific community is the term of "biomarkers". In the following sections the use of mussels for assessing xenobiotic exposure and effects at different levels of organization will be described.

2.1. Exposure Biomonitoring

The major sources of water pollution are chemical, physical or biological.

2.1.1. Chemical Contamination

2.1.1.1. Metal Biomonitoring

Anthropogenic sources of metals are mainly industrial, petroleum contamination and sewage disposal. Metal ions are concentrated in aquatic organisms and move up the food chain. Some of them, the heavy metals like cadmium, zinc, lithium and aluminium have been released to the environment with the industrial development. These metals can have drastic embryotoxic/teratogenic effects on organisms (Calevro et al., 1998), whereas other metals such as Zn, Cu and Fe at trace levels are beneficial to organisms because they play a crucial role in the life processes.

Low concentration of metals in water makes the analytical analysis very difficult. The metal concentration in the sediment can provide an estimate but the measured levels are significantly affected by the particle size, organic content and red-ox conditions. Biomonitors, such as mussels, can concentrate the metals from the water and also provide a time-integrated estimate on the bioavailable fraction of metals in the environment (Wu et al., 2007). Prior knowledge of the accumulation kinetics of a particular metal in a particular organism must be achieved to convert the accumulated metal concentration into a measure of all the bioavailable supplies of that metal to the organism over a time period. Mussels usually filter suspended detritus particles and small size phytoplankton, which are more likely to be organically and metal rich particles than the zooplankton counterparts (Rainbow, 2006). Therefore several studies on different species of mussels, have been used for biomonitoring metals in the "mussel watch" program (Cantillo, 1998; Goldberg, 1975; Goldberg et al., 1978; Hunt & Slone, 2010; Lauenstein et al., 1990; Phillips, 1978; Richardson et al., 1994) and in other studies (Besada et al., 2011; Boening, 1999; Gundacker, 1999; Hendriks et al., 1998; Johns & Timmerman; 1998, Julshamn & Grahl-Nielsen, 1996; Phillips & Rainbow, 1993; Rainbow, 2006; Rainbow & Phillips 1993; Richman & Somers, 2005; Richman & Somers, 2010; Riget et al., 1996; Szefer et al. 2006; Sunlu, 2006; Usero et al., 1996; Yap et al., 2006).

2.1.1.2. Pesticides and Herbicides Biomonitoring

Many water sources including drinking supplies are contaminated with pesticides/herbicides from widespread agricultural uses. These chemicals are readily available to wild animals and even more worryingly they are passed up on the food chain. A major characteristic of organic pollutants is their persistence in the environment due to their ability to accumulate and concentrate in biological systems (IEH, 1995). In particular dichlorodiphenyldichloroethane (DDT), a polychlorinated pesticide that is resistant to destruction by light and oxidation, and widely used in 1940s and 1960s to control insect vectors of malaria, yellow fever, sleeping sickness and river blindness. Banned in the west in 1970s, because of its ability to cause endocrine disruption and act as a cancer promoter (Burlington & Lindeman, 1950; Wolf & Toniolo, 1995), it is still used as a cheap alternative to fight against malaria in some developing countries. Biomonitoring programmes tracing DDT and its metabolites successfully employ different mussel species all over world: zebra mussel, *D. polymorpha*, (Baccetta & Mantecca, 2009; Binelli & Provini, 2003; Binelli et al., 2001; Hendriks et al., 1998), the blue mussel, *Mytilus edulis* (Gustavson & Jonsson, 1999; O' Connor, 1996; Thompson et al, 1999), the freshwater mussels, *E. complanata* and *Lampsilis radiata radiata* (Metcalfe & Charlton, 1990, the scorched mussel, *Brachidontes exustus* (Tavares et al., 1999) or the Asian green mussel, *Perna viridis* (Liu & Kueh, 2005; Tanabe et al., 2000).

Lindane is another persistent environmental pollutant, manufactured as hexachlorocyclohexane since the 1940s. This organochlorine pesticide was even used as the active ingredient in several prescribed head lice repellent shampoos. In 2009 the UN banned its production and use worldwide because of its neurotoxic effects. Lindane has been detected not only in agricultural head-water streams but also in wastewater treatment plant effluents (Humphreys et al., 2008). Zebra mussel (*D. polymorpha*), the Mediterranean blue mussel (*Mytilus galloprovincialis*) and the green mussel (*P. viridis*) have been used as indicators of water contamination with lindane (Berny et al., 2002; Boonyatumanond et al., 2002; Hendriks et al., 1998).

To date, the development of imposex in gastropods exposed to TBT is considered one of the clearest examples of chemically induced endocrine disruption (Gibbs & Bryan, 1986). The use of organotin-based paints on vessels less than 25m in length was banned for most of the countries in 1987 and since 2003 in all ships. Despite such restrictions, TBT persists in many areas at high levels (Berto et al., 2007). Organotin compounds have been biomonitored using mussels such as *M. galloprovincialis* (Barroso et al., 2004; Ruiz et al., 2005), *M. edulis* (Albalat et al., 2002; Følsvik N, et al., 2002), *D. polymorpha* (Roper et al., 2001; Stäb et al., 1995) or *P. virdis* (Sudaryanto et al., 2002; Tanabe et al., 2000).

Very few studies have focused on the analysis of other organochlorine pesticides such as heptachlor, chlordane and toxaphene, in species such as *M. edulis* (Pineiro et al., 1995), *P. viridis* (Boonyatumanond et al., 2002) and *D. polymorpha* (Hendriks et al., 1998).

2.1.1.3. Other Industrial-Derived Chemicals Biomonitoring

Polychlorinated biphenyl (PCBs) and alkylphenolic compounds (APEs) are toxic industrial-derived chemicals present in aquatic environments. PCBs have been produced since 1929 for various purposes such as cooling and hydraulic liquids, dielectric fluids in transformers, flame-retardants and as components of plastics and paints. They are still present in nearly every environmental compartment or matrix, almost 30 years after their ban

(Lintelmann et al., 2003). Biomonitoring of PCBs is also performed using mussels such as *E. complanata* (Drouillarda, et al., 2013; Gewurtz et al., 2003; Metcalfe & Charlton, 1990), *M. edulis* (Hunt & Slone, 2010), *D. polymorpha* or *Dreissena bugensis* (Richman & Somers, 2005) or *P. viridis* (Liu & Kueh, 2005)

APEs are used as non-ionic surfactants, plastic additives, emulsifiers, and wetting agents in agricultural and industrial applications, as well as spermicides in contraceptive applications. In wastewater treatment plants and in the environment, APEs can be microbially degraded, but toxic and estrogenic metabolites such as *p*-nonylphenols and *p*-nonylphenol ethoxylates are formed (Lintelmann et al., 2003). The occurrence of APEs in the aquatic environment can be biomonitored using *E. complanata* (Sabik et al., 2003) or *M. galloprovincialis* (Ferrara et al., 2001).

A number of other compounds such as bisphenol A (from polycarbonate flasks), polycyclic aromatic hydrocarbons (PAHs generated during the combustion of organic matter), polychlorinated dibenzo dioxins (tetrachlorodibenzo-*p*-dioxin TCDD and –furans by-products of industrial processes) are also environmental contaminants and have been shown to disrupt the endocrine system of different organisms (IEH, 1995; Lintelmann et al., 2003). Different species of mussels are successfully used to monitor these contaminants all over world (Gewurtz et al., 2003; Hunt & Slone, 2010; Isobe et al., 2007; Kaus & Hamdy, 1991; Liu & Kueh, 2005).

2.1.2. Physical Contamination

Nuclear pollution accounts for only a small amount of oceanic pollution. Some of the nuclear exposure comes from radioactive substances (radionuclides) that occur naturally. Occasionally these substances can accumulate through human activities such as mining, nuclear energy production, and weapons testing resulting in a higher exposure. While significant amounts of radium, plutonium, and other radioactive materials can cause ecological damage in isolated areas, such as a bay neighbouring a leaking radioactive materials depository, the ocean, as a whole, remains relatively unaffected by the global increase in nuclear materials (Lionetto et al., 2004). The protection from nuclear accidents and the management of radioactive wastes for the protection of humans and the environment are important elements of radiation protection policies.

To determine the present status of radionuclide exposure in the coastal and estuarine environments also mussels such as *Mytilus trossulus*, *M. galloprovincialis, M. edulis*, *Mytilus californianus* or *D. polymorpha* are used as biomonitors (some of these studies are part of the "mussel watch" monitoring program) (Charmasson et al., 1999; CIESM, 2002; Goldberg et al, 1983; Mersch et al., 1992; Thébault, 2005; Valette-Silver & Lauenstein, 1995).

2.1.3. Biological Contamination

The presence of bacteria in aquatic environments can have important human health consequences, as well as ecological and economic implications. The levels of bacteria have to be monitored and have to comply with the allowable concentrations set by the local regulations and when needed the sources of bacterial contamination have to be eliminated. Therefore it is important to develop an efficient monitoring method for locating bacterial sources. In this frame, *D. polymorpha, M. edulis* and *Perna perna* have been assessed for their capability to accumulate thermotolerant coliforms (mainly *Escherichia coli*) and *Enterococcus* for faecal pollution monitoring (Igawa Martinez & Fernandes Cardoso de

Oliveira, 2010; Rosley et al., 2009; Selegean et al., 2001). The use of mussels for bacterial monitoring however is limited because the relationship between the numbers of bacteria accumulated in the mussel with that found in the water column is very difficult to evaluate. This is due to complex interactions related to the physiology of the bivalve and the bacterial characteristics, which can also be affected by several environmental conditions (Sinton et al., 2002; Solé et al., 2010).

2.2. Effects Biomonitoring

2.2.1. Biochemical Indicators

Alterations in different biochemical parameters can represent specific markers of exposure and/or biological effects to a particular class of contaminant. When referred to individual organism these responses are often referred as biomarkers. The use of biomarkers, have become a common practice as "early warning" pollution monitoring tools. The following sections briefly describe a number of common cellular and biochemical biomarkers of marine pollution used in mussels.

2.2.1.1. Molecular, Cellular and Tissue Parameters

Exposure of mussels to environmental contaminants can cause DNA damage, such as chromosomal aberrations, sister-chromatid exchange, micronuclei formation and strand breaks (Vukmirović et al., 1994). In *M. galloprovincialis*, DNA-single strand break and micronuclei frequencies are sensitive biomarkers for monitoring heavy metal contamination and other environmental pollutants (Bolognesi et al., 1999; Scarpato et al., 1990). A micronuclei cytome assay has been developed in mussels for measuring chromosomal DNA damage as a result of either chromosome breakage or chromosome missegregation during mitosis (Bolognesi & Fenech, 2012). This test can be widely used in biomonitoring screening for potential genotoxic compounds. Single strand breaks are associated with urban and industrial contaminated waters (Vukmirović et al., 1994). Biomonitoring of genotoxic contamination using the comet assay to evaluate strand breaks has been successfully employed in mussels (Alamri et al., 2010; Villela, 2007). DNA-protein and DNA-DNA crosslinks have been used as biomarkers of exposure to urban and industrial waste and to benzo[a] pyrene in *M. galloprovincialis* (Vukmirović et al., 1994; Canova et al., 1998). Chromosomal aberrations (counted as break and fragments) are also considered a relevant parameter to assess exposure to genotoxic chemicals in *M. galloprovinciallis* (Al-Sabti & Kurelec, 1985).

A cytogenic endpoint, the sister chromatid exchange, has been also used to assess chemically induced genetic damage in *Mytilus* species but its temporal and spatial variations poses difficulties to be applied directly to the field (Dixon & Clarke, 1982; Pasantes et al., 1996). In *M. galloprovincialis*, nuclear buds are associated with metal contamination (Fernández et al., 2011) whereas binucleated cells appear to be more correlated to the accumulation of organic pollutants. The appearance of these nuclear abnormalities seems to be linked to temperature and therefore should be taken into account on the application of these biomarkers as pollution screening tools (Fernández et al., 2011).

Lysosomes in the mussel digestive gland take and accumulate metals and organic xenobiotics and play a role in detoxification processes. Accumulation of these compounds and their metabolic products impair the stability of lysosomal membrane. Therefore, lysosomal alterations can be used as a biomarker of contaminated environments in mussels (Cajaraville et al., 1991; Petrović et al., 2001; Regoli et al., 1992).

Tissue alterations in selected organs and tissues, mainly gills, liver and gonad, are considered as histopathological or tissue-level biomarkers in mussels exposed to environmental contaminants (Au, 2004).

2.2.1.2. Alteration of Biochemical Pathways and Enzyme Activities

Metabolism of lipophylic xenobiotics in organisms occurs in two phases. Phase I metabolism (functionalisation) involves oxidation, hydrolysis, hydration or reduction and biotransforms the parent compound into metabolites of greater polarity. Phase II metabolism involves conjugation to produce polar conjugates that can be readily excreted by the organism. In mussels, Phase I reactions take place mainly in the digestive gland which is the major site of biotransformation.

During Phase I metabolism the xenobiotic can experience a series of oxidative processes mediated by the cythocrome P450-dependent monooxygenase (terminal component of the mixed-function oxygenase system) with the aim of producing more polar metabolites. Induction of cytochrome P4501A (CYP1A) is routinely used in fish as a specific biomarker of exposure to organic contaminants. However this is applied to a much less extent in molluscs due to the lower apparent induction and higher variability of the system than in fish (Porte et al., 2001; Solé et al., 2000). Preliminary studies on the use other CYP families involved in the metabolism of xenobiotics for biomonitoring pollution has been carried out (Cubero-Leon et al., 2012; Jonsson et al., 2004). Ethoxyresorufin-*O*-deethylase (EROD) activity assay is a highly sensitive biomarker of contaminant uptake in fish. It measures the oxidative deethylation of 7-ethoxyresorufin to resorufin, catalyzedby cytochrome P-450 CYP 1A1, which is the principal P-450 isozyme involved in the *O*-deethylation of the substrate. Therefore, the EROD assay is widely used in vivo and in vitro as a marker for exposure to CYP 1A1 inducers such as exposure to certain planar halogenated and polycyclic aromatic hydrocarbons. In mussels living in a lake contaminated with DDT, PCBs, PAHs and heavy metals EROD activity is activated (Binelli et al., 2005).

The phase I by-products may undergo further metabolism by conjugative enzymes (phase II) such as glutathione S-transferase (GST). In mussels GST is used as a biomonitoring tool of aquatic contamination (Sáenz et al., 2010; Serafim et al., 2011). These reactions might bring the synthesis of more reactive molecules that can interact with the genetic material and cause DNA damage (see previous section). They may also increase the presence of reactive oxygen species (ROS). In order to prevent the formation of ROS and to protect the organism from oxidative stress, antioxidant enzymes such as catalase, superoxide dismutase, glutathione peroxidise and DT-diaphorases are involved. These enzymes are used as oxidative stress biomarkers in mussels exposed to different types of pollutants (Serafim et al., 2011; Solé et al. 2000). Peroxisomes are cell organelles that play key roles in multiple cell functions especially in the metabolism of ROS. In mussels, peroxisomes proliferate under exposure to certain xenobiotics such as PAHs, crude and lubricant oils, PCBs and phthalate plasticizers (Cajaraville et al., 2003). The peroxisome proliferation in digestive gland can be measured in terms of acyl-CoA oxidase (AOX) activity as exposure biomarker in mussels (Marigómez et al., 2006).

Contamination of heavy metals can be examined by tracking changes in metallothionein (MT), a well-known biomarker involved in the sequestration and detoxification of metals (Banni, et al., 2007; Serafim et al., 2011). MTs are a family of cysteine-rich proteins that have the capacity to bind to heavy metals through the thiol group of its cysteine residues. Although MT function is not clear it appears to provide protection against metal toxicity and oxidative stress.

Organic pollutants exert neurotoxic effects and they can be biomonitored using acetylcholinesterase (AChE) activity. AChEs are located at neuromuscular junctions in the central and peripheral nervous system. Their function is to regulate the nervous transmission by reducing the concentration of ACh in the junction by hydrolysis. Cholinesterases can also be found in other organs although the physiological functions of AChE in tissues other than nerves are not known (Rickwood & Galloway, 2004). The inhibition of AChE activity has been widely used as a biomarker of exposure to pesticides in mussels (Mora et al., 1999; Rickwood & Galloway, 2004).

Xenobiotics can also have reproductive effects. Exposure to endocrine disruptors (EDCs) has occurred in aquatic environments worldwide and there is sufficient evidence to conclude that adverse endocrine-mediated effects have occurred in some species (Gibbs & Bryan, 1986; Jobling et al., 1998). In vertebrate species, production of vitellogenin (egg yolk precursor) by male fish and an increase in the incidence and severity of intersex (testes containing eggs) (Jobling et al, 1998) has been associated with the presence of estrogenic chemicals in treated domestic sewage effluents. However, the risk of estrogenic chemicals to aquatic invertebrates is less clear and therefore the use of reproductive end-points as biomarkers of estrogenic pollution in mussels have not yet been established. Exposure of adult mussels to EDs such as crude oil and produced water mixtures found in offshore oil extraction platforms can provoke changes in phosphoproteins and they can be measured using alkali-labile phosphate quantification (vitellogenin-like protein levels) (Ortiz-Zarragoitia & Cajaraville, 2006). It has also been suggested that the reproductive physiology of molluscs, in terms of, vitellogenin, estrogen receptor and serotonin receptor mRNA expression, may be susceptible to damage by environmental estrogens at certain points in their gametogenesis process (Ciocan et al., 2010; Cubero-Leon et al., 2010).

The main function of the immune system for all metazoan organisms is to sustain host defence mechanisms and maintain homeostasis within the organism. In bivalves, phagocytic activity, considered as an indicator of immunotoxicity, can be induced by a wide range of xenobiotics (Blaise et al., 2002). In mussels, immunological biomarkers including lysozyme, phenoloxidase, and phagocytic activities or phagocytosis have been applied in monitoring contaminated sites (Pipe et al., 1995; Yaqin, 2011).

2.2.1.3. Multimarker Approaches

These biochemical indicators can be useful to scientists to evaluate specific responses to natural or anthropogenic changes, but it is very difficult for environmental managers to interpret increasing or decreasing changes in biomarker data with water pollution (Narbonne et al., 2001). This is why nowadays; multimarker measurements are used as a diagnostic tool of pollution status.

Some multimarker approaches involve specific statistical analysis for pattern discrimination between the different locations in relation with their pollution levels as evaluated by chemical measurements. Then, calculation of pollution indexes based on a specific scale taken into account multimarker response should be able to translate scientific data into useful information to estimate the quality level of coastal environments (Narbonne et al., 2001). In addition, integrated biomarker response (IBR) in mussels is a method appears to be a useful tool for scientists and managers in assessing ecological risk. This is a simple method to summarize biomarker responses and simplify their interpretation in biomonitoring programs (Beliaeff & Burgeot, 2002; Damiens et al., 2007). It uses star plots to display results for the panel of biomarkers used for each station and survey. IBR is then computed as the star plot area. Star plots using IBR values instead of biomarker data make it possible to visualize between-site and/or between-survey differences for comparison with exposure conditions. Viarengo et al., (2007) have also proposed a 2 tier approach for assessing the level of pollutant-induced stress syndrome in mussels. This involves two steps. An initial screening (Tier 1) with a low-cost biomarker such as lysosomal membrane stability and a Tier 2 step in sites where an alteration in lysosomal stability has been established. Tier 2 consists in assessing levels of stress syndrome by using a full battery of makers. The data collected should be then run through an appropriate approach of data integration to obtain an objective assessment of the pollutant induced stress syndrome.

2.2.2. Physiological and Behavioural Indicators

Mytilus sp have been widely studied as bio-indicator species. The ubiquitous and sessile adult condition is particularly challenged by the dynamic habitat they populate. Among the most important are periodic extreme variations in body temperature, especially when mid-day low tides coincide with high solar radiation.

It is well accepted today that environmental changes can trigger off compensatory reactions in mussels, such as valve closure. This simple behaviour may have important ecological and physiological implications. Sealing of the valves represents a primary anti-predator defence mechanism (Robson et al., 2010), especially against starfish, yet valve closure may have negative consequences for the organism' entire metabolism. When valves are sealed closed, the individual no longer has access to particulated food and oxygenated seawater. Metabolism shifts to less efficient anaerobic pathways and total metabolic output consequently decreases (Dowd & Somero, 2012). It has been suggested that valve movement and shell gaping and other physiological processes such as byssal attachment or growth in freshwater and marine mussels can be used for monitoring changes in the environmental quality (Kramer et al., 1989). These can be exposure to trace metals (Manley & Davenport 1979), pesticides (Köprücü et al., 2010; Sálanki & Varanka, 1978) and in general to detect arising toxic conditions in the water (Slooff et al., 1983). Byssogensis test can also offer a rapid technique for monitoring exposure to sublethal concentrations of a wide range of potentially toxic environmental chemicals, such as heavy metals or pesticides (Roberts, 1975).

On the other hand, opening of the valves ('gaping') during episodes of elevated body temperature is probably used to ameliorate the effects of high temperature via evaporative cooling, by maintaining mantle cavity fluid oxygenation to support aerobic metabolism. Mussels' modulation of the valve gape in response to thermal extremes and its physiological consequences may have profound effects on the survival, distribution and abundance of the species (Dowd & Somero 2012) which can be a result of environmental global change and may be used to monitor it. However, much work remains to be done to be able to use this behaviour as routine biomonitoring tool for environemtal change (Zippay & Helmut, 2012).

Other physiological tests have been used to explain and predict the success of newly introduced species, which have invaded a new environment accidentally or on purpose. Braby and Somero (2006) used *Mytilus* congeners: *M. trossulus,* and *M. galloprovincialis* to study their distinct global pattern distribution. Physiological experiments showed a significantly higher heart rate for *M. Trossulus* than that of *M. Galloprovincialis,* consistent with evolutionary adaptation to a lower habitat temperature in the former species (temperature compensation). Mussels' heart rate modulations are used in oil pollution monitoring. Halldórsson et al., (2008) noted that mussels from the polluted site (Reykjavík harbour) increased their heart rate when exposed to benzo[a]pyrene, while no difference was observed between the heart rate values of the individuals from the clean site.

A body weight to length ratio, calculated as a body condition index (BCI) has been intensively used in biomonitoring studies for the past 30 years. Various species of mussels are collected from suspected sites and health is assessed using: soft tissue, mantle, total gill and shell wet weights; shell length, shell width. Overall, these data can indicate that both habitat degradation and elevated contaminant levels from anthropogenic sources are probable causes of mussel poor BCI (Riget et al., 1996).

Ruessler et al., (2011) examined species diversity, health, and reproductive status of freshwater mussels in South Florida in relation to habitat alteration and contaminants. Their study recorded the decline of populations, most probably due to the approximately 2 million acres of the Everglades drainage system, containing sediments and biota with elevated mercury and other contaminants related to human activities.

2.2.3. Population, Communitry Parameters and Ecosystem Effects

Pollution and overfishing are the main concerns in marine ecosystems nowadays. Anthropogenic introduced xenobiotics may affect genetic diversity at population levels and they may also alter the genetic structure of the population with catastrophic consequences. This reduction in genetic diversity results from a selection for chemical tolerance, increased mortality and reduction in the number of breeding adults (Fernandez-Tajes, 2012).

Mussel populations are deeply affected by long time exposure to various marine pollutants and they are intensively used to assess ecological impact of environmental contaminants on marine communities. Indicators such as diversity index or population densities have been recently employed to examine the temporal pattern of genetic distribution among populations of mussels, following the Prestige oil spillage, in Spain (Fernandez-Tajes, 2012). Catastrophic events, such as Prestige, have a huge effect on low mobility or sessile marine organisms, which suffer more acutely the toxic consequences of the oil spill.

Heterozygosity values among mussel populations reflect the relationship between genetic and environmental factors. It determines the fitness of the population since heterozygosity increases adaptability. Bivalves usually live in massive populations with a high immigration rate (due to pelagic larvae stages) and these facts should keep a high heterozygosity value (Arias et. al, 2010). Nevertheless, oil compounds, most of them considered carcinogenic, would lead to changes in allelic frequencies, mainly due to sub lethal effects of PAHs and PCBs on gametogenesis.

Reduced reproductive success, deficiencies of gonad development coupled with low rate of larvae survival and changes in body indexes of breeding adults are together contributing to a reduction in genetic diversity of mussel populations, similar to "sweepstakes hypothesis" (Hedgecock, 1994). The assessment of genetic diversity of a population, especially mussel populations, which occupy an important position the marine food web, is considered a valuable ecological indicator for determining the effects of pollution in aquatic ecosystems.

The use of caged mussels can easily become very informative when quantifying the number of mussels surviving a 30 day-caging period in polluted waters and survival time when exposing animals to air (Viarengo et al., 2007). The second biomarker can be employed to assess the effects of environmental stressors at the organism level to clearly determine whether contaminants have affected the capacity of molluscs to survive to further environmental change (e.g. air exposure) (Moles & Hale, 2003).

In recent years, the toxic effects caused by endocrine disrupting chemicals on the eco-environment have interest from scientists all over the world. TBT (tributyltin) was artificially introduced to marine environment, and has posed a lot of threat to marine aquatic organisms. Some marine gastropod species have been profoundly affected by imposex and the effects have been recorded at higher organizational levels: decline of gastropod population, the damage of the marine food chain, and, as time passed, TBT did serious damage to the eco-environment (Gibbs & Bryan, 1986).

Conclusion and Perspectives

Great progress has been achieved in the field of biomonitoring of environmental pollution due to the efforts of multidisciplinary and international biomonitoring programs.

Mussels have been used as sentinel organisms to biomonitor a variety of chemicals in the aquatic environment such as metals, organochlorines, pesticides and PAHs and are being used in biomonitoring studies from international programmes like "Mussel Watch" (Goldberg, 1975). Their sessile nature, wide geographical distribution, large populations, high filtering rates, and poor ability to metabolise xenobiotics make them ideal organisms for bioaccumulating contaminants from the aquatic environment. In recent years, there has been increasing public and scientific concern on the occurrence and contamination by emerging pollutants. The EPA (Environmental Protection Agency-United States) defines emerging pollutants as new chemicals without regulatory status and which impact on environment and human health are poorly understood. Currently, the potential use of mussels to elucidate the occurrence of some of these pollutants in the environment, such as pharmaceuticals and personal care product, is being under study (Nakata et al., 2012). However, little information on ecotoxicological implications is still available (Ciocan et al., 2010; Pal et al., 2010) and further monitoring studies are needed to assess the trends and risks associated with their accumulation in the aquatic environment.

A variety of biochemical and physiological tests are being used nowadays as indicators and perhaps eventually as predictors of the effects of toxicants. They are known by the scientific community as biomarkers. The biomarker approach can be a useful means of detecting exposure *in situ,* providing early indications of ecologically relevant effects. Nowadays, a lot of effort has been put into developing new biomarkers of exposure and effects in mussels. For example changes in cellular signalling or different mRNA level changes to monitor pollutant effects are currently being studied (Viarengo et al., 2007; Ciocan et al., 2010; Cubero-Leon et al., 2010, Cubero-Leon et al., 2012). Research approaches such as proteomics and metabolomics are now starting to be used for studying and understanding the effects of environmental pollutants in molluscs (Apraiz et al., 2006; Kwon et al., 2012). The application of these "omics" technologies could provide a global assessment of the general status of an organism, increasing our understanding of the mode of action of certain pollutants and their potential toxicity. It should be noted that their application could also lead to the discovery of new families of biomarkers.

There is also a need for an integrative tool and higher harmonization criteria among regional programs to compare results. Future goals include the use of these tools to be able to estimate risk and to summarize the quality of the water to be used by decision-makers (Cajaraville et al., 2007).

In conclusion, even though there are a high number of research studies for biomonitoring environmental pollution and for the development of new biomonitoring tools, further development in this area is urgently needed. There are still many questions to answer about presently used biomarkers, their relation to health effects and their ability to be used by decision-makers. A deeper understanding of sentinel organism physiology is needed to correctly interpret the responses observed.

References

Alamri, O.D., Cundy, A.B., Di, Y., Jha, A.N. & Rotchell, J.M. (2012). Ionizing radiation-induced DNA damage response identified in marine mussels, *Mytilus* sp. *Environmental Pollution*, 168, 107-112.

Albalat, A., Potrykus, J., Pempkowiak, J. & Porte C. (2002). Assessment of organotin pollution along the Polish coast (Baltic Sea) by using mussels and fish as sentinel organisms. *Chemosphere*, 47, 165-171.

Al-Sabti, K. & Kurelec, B. (1985). Environmental contamination land toxicology induction of chromosomal aberrations in mussel *Mytilus galloprovincialis* Watch. *Bulletin of Environmental Contamination and Toxicology*, 35, 660-665.

Apraiz, I., Mi, J. & Cristobal, S. (2006). Identification of proteomic signatures of exposure to marine pollutants in mussels (Mytilus edulis). *Molecular and Cellular Proteomics*, 5, 1274-1285.

Arias A., Fernandez-Moreno, M., Fernandez-Tajes, J., Gaspar, M.B. & Mendez, J. (2010). Strong genetic differentiation among east Atlantic populatioins of the sword razor shell (*Ensis siliqua)* assessed with mtDNA and RAPD marquers. *Helgolandd Marine Research*, 65, 81-89.

Au, D.W.T. (2004). The application of histo-cytophathological biomarkers in marine pollution monitoring: a review. *Marine Pollution Bulletin*, 48, 817-834.

Bacchetta, R. & Mantecca, P. (2009). DDT polluted meltwater affects reproduction in the mussel *Dreissena polymorpha*. *Chemosphere*, 76, 1380-1385.

Banni, M., Dondero, F., Jebali, J., Guerbej, H., Boussetta, H. & Viarengo, A. (2007). Assessment of heavy metal contamination using real-time PCR analysis of mussel metallothionein mt10 and mt20 expression: a validation along the Tunisian coast. *Biomarkers*, 12, 369-383.

Barroso, C.M., Mendo, S. & Moreira, M.H. (2004). Organotin contamination in the mussel *Mytilus galloprovincialis* from Portuguese coastal waters. *Marine Pollution Bulletin*, 48, 1145-1167.

Beliaeff, B. & Burgeot, T. (2002). Integrated biomarker response: a useful tool for ecological risk assessment. *Environmental Toxicology and Chemistry*, 21, 1316-1322.

Berny, P., Lachaux, O., Buronfosse, T., Mazallon, M. & Gillet, C. (2002). Zebra mussels (*Dreissena polymorpha*) as indicators of freshwater contamination with lindane. *Environmental Research*, 90, 142-151.

Berto, D., Giani, M., Boscolo, R., Covelli, S., Giovanardi, O., Massironi, M. & Grassia, L. (2007). Organotins (TBT and DBT) in water, sediments, and gastropods of the southern Venice lagoon (Italy). *Marine Pollution Bulletin,* 55, 425-435.

Besada, V., Andrade, J.M., Schultze, F. & Gonzalez J.J. (2011). Monitoring of heavy metals in wild mussels (*Mytilus galloprovincialis*) from the Spanish North-Atlantic coast. *Continental Shelf Research*, 31, 457–65.

Binelli, A., Galassi, S. & Provini, A., (2001). Factors affecting the use of *Dreissena polymorpha* as a bioindicator: the PCB pollution in Lake Como (N. Italy). *Water Air and Soil Pollution*, 125, 19–32.

Binelli, A. & Provini, A. (2003). DDT is still a problem in developed countries: the heavy pollution of Lake Maggiore. Chemosphere 52, 717–723.

Binelli, A., Ricciardi, F., Riva, C. & Provini, A. (2005). Screening of POP pollution by AChE and EROD activities in Zebra mussels from the Italian Great Lakes. *Chemosphere*, 61, 1074-1082,

Blaise, C., Trottier, S., Gagné, F., Lallement, C. & Hansen, P.D. (2002). Immunocompetence of bivalve hemocytes as evaluated by a miniaturized phagocytosis assay. *Environmental Toxicology*, 17, 160-169.

Boening, D.W. (1999). An evaluation of bivalves as biomonitors of heavy metals pollution in marine waters. *Environmental Monitoring and Assessment*, 55, 459-470.

Bolognesi, C. & Fenech, M. (2012). *Nature Protocols*. 7, 1125-1137.

Bolognesi, C., Landini, E., Roggieri, P, Fabbri, R & Viarengo, A. (1999). Genotoxicity biomarkers in the assessment of heavy metal effects in mussels: experimental studies. *Environmental and Molecular Mutagenesis*, 33, 287-292.

Boonyatumanond, R., Jaksakul, A., Puncharoen, P. & Tabucanon, M.S. (2002). Monitoring of organochlorine pesticides residues in green mussels (*Perna viridis*) from the coastal area of Thailand. *Envrionmental Pollution*, 119, 245-252.

Braby, C.E., & Somero, G.N. (2006). Ecological gradients and relative abundance of native (Mytilus *trossulus*) and invasive (*M. galloprovincialis*) blue mussels in the California hybrid zone. *Marine Biology*, 148, 1249-1262.

Burlington, H. & Lindeman, V.F. (1950). Effect of DDT on testes and secondary sex characters of white leghorn cockerels. *Proceedings of the Society for Experimental Biology and Medicine*, 74, 48-51.

Cajaraville, M.P., Cancio, I., Ibabe, A. & Orbea, A. (2003). Peroxisome proliferation as biomarker in environmental pollution assessment. *Microscopy Research and Technique*, 611 191-202.

Cajaraville, M.P., Marigomez, J.A. & Angulo, E. (1991). Automated measurement of lysosomal structure alterations in oocytes of mussels exposed to petroleum hydrocarbons. *Archives of Environmental Contamination and Toxicology,* 21, 395-400.

Cajaraville, M.P, Theron, M. & Le Floch, S. (2007). A Pragmatic and Integrated Approach for the Evaluation of Environmental Impact of Oil and Chemical Spills at sea: Input to European Guidelines (PRAGMA). Tehcnical Report. http://ec.europa.eu/echo/files /policies/marine_pollution/techrep.pdf

Calevro, F., Campani, S., Ragghianti, M., Bucci, S. & Mancino, G. (1998). Test of toxicity and teratogenicity in biphasic vertebrates treated with heavy metals (Cr^{3+}, Al^{3+}, Cd^{2+}). *Chemosphere*, 37, 3011-3017.

Cantillo, A.Y. (1998). Comparison of results of Mussel Watch programs of the United States and France with worldwide Mussel Watch studies. *Marine Pollution Bulletin*, 36, 712-717.

Canova, S., Degan, P., Peters, L.D., Livingstone, D.R., Voltan, R. & Venier P. (1998). Tissue dose, DNA adducts, oxidative DNA damage and CYP1A-immunopositive proteins in mussels exposed to waterborne benzo[a]pyrene. *Mutational Research*, 399, 17-30.

Carson, R. (1963). Silent spring. New York: Houghton Mifflin.

Charmasson, S., Barker, E., Calmet, D., Pruchon, A.S. & Thébault H. (1999). Long-term variations of man-made radionuclide concentrations in a bioindicator *Mytilus galloprovincialis* from the French Mediterranean coast. *Science of the Total Emvironment,* 30, 93-103.

CIESM, (2002). Mediterranean mussel watch – designing a regional program for detecting radionuclides and trace-contaminants. In: CIESM Workshop Monographs, 15. http://www.ciesm.org/online/monographs/Marseilles.pdf

Ciocan, C.M., Cubero-Leon, E., Puinean, A.M., Hill, E.M., Minier, C., Osada, M., Fenlon, K. & Rotchell, J.M.(2010). Effects of estrogen exposure in mussels, Mytilus edulis, at different stages of gametogenesis. *Environmental Pollution*, 158, 2977-2984.

Cubero-Leon, E., Ciocan, C.M., Hill, E.M., Osada, M., Kishida, M., Itoh, N., Kondo, R., Minier, C. & Rotchell, J.M. (2010). Estrogens disrupt serotonin receptor and cyclooxygenase mRNA expression in the gonads of mussels (*Mytilus edulis*). *Aquatic Toxicology*, 98, 178-187.

Cubero-Leon, E., Puinean, A.M., Labadie, P., Ciocan, C.M., Naoki, Itoh, N., Kishida, M., Osada, M., Minier, C., Hill, E.M. & Rotchell, J.M. (2012). Two *CYP3A-like* genes in the marine mussel *Mytilus edulis*: mRNA expression modulation following short-term exposure to endocrine disruptors. *Marine Environmental Research*, 74, 32-39.

Damiens, G., Gnassia-Barelli, M., Loquès, F., Roméo, M. & Salbert, V. (2007). Integrated biomarker response index as a useful tool for environmental assessment evaluated using transplanted mussels. *Chemosphere*, 66, 574-583.

Dixon, D.R. & Clarke, K.R. (1982). Sister chromatid exchange: a animal in genetic toxicology. *Mutational Research*, 97,35-42.

Dowd, W.W. & Somero, G.N. (2012). Behavior and survival of *Mytilus* congeners following episodes of elevated body temperature in air and seawater. *The Journal of Experimental Biology*, [Epub ahead of print].

Drouillard, K.G., Jezdic, I., O'Rourke, S.M., Gewurtz, S.B., Raeside, A.A., Leadley, T.A., Drca, P. & Haffner, G.D. (2013). Spatial and temportal variability of PCBs in Detroit

River water assessed using a long term biomonitoring program. *Chemosphere*, 90, 95-102.

Fernández, B., Campillo, J.A, Martínez-Gómez, C. & Benedicto, J. (2011). Micronuclei and other nuclear abnormalities in mussels (Mytilus galloprovincialis) as biomarkers of cyto-genotoxic pollution in mediterranean waters. *Environmental and Molecular Mutagenesis,* 52, 479-491.

Fernández-Tajes, J., A. Arias-Pérez, A., Fernández-Moreno, M. & Méndez, J. (2012). Sharp decrease of genetic variation in two Spanish localities of razo clam *Ensis siliqua*: natural fluctuation or Prestige oil spill effects? *Ecotoxicology*, 21, 225-233.

Ferrara, F., Fabietti, F., Delise, M., Bocca, A.P. & Funari, E. (2001). Alkylphenolic compounds in edible molluscs of the Adriatic Sea (Italy).*Environmental Science and Technology*, 35, 3109-3112.

Følsvik, N., Brevik, E.M. & Berge, J.A. (2002). Organotin compounds in a Norwegian fjord. A comparison of concentration levels in semipermeable membrane devices (SPMDs), blue mussels (*Mytilus edulis*) and water samples. *Journal of Environmental Monitoring*, 4, 280-283.

Gewurtz, S.B., Lazr R. & Haffner, G.D. (2003). Biomonitoring of bioavailable PAH and PCB water concentrations in the Detroit River using the freshwater mussel, *Elliptio complanata*. *Journal of Great Lakes Research*, 29, 242-256.

Gibbs, P. E. & Bryan, G. W. (1986). Reproductive failure in populations of the dogwhelk, *Nucella lapillus,* caused by imposex induced by tributyltin from antifouling paints. Journal of the Marine Biological Association of the United Kingdom, 66, 767-&.

Goldberg, E.D. (1975). The mussel watch-a first step in global marine monitoring. *Marine Pollution Bulletin*, 6, 111-111.

Goldberg, E.D., Bowen, J.W., Robertson, E., Schneider, E. & Gamble, E. (1978). The Mussel Watch. *Environmental Conservation*, 5, 101-125.

Goldberg, E.D., Koide, M., Hodge, V., Flegal, A.R. & Martin, J. (1983). U.S. mussel watch: 1977-1978 results on trace metals and radionuclides. *Estuarine, Coastal and Shelfish Science*, 16,69-93.

Gundacker, C. (1999). Tissue-specific heavy metal (Cd, Pb, Cu, Zn) deposition in a natural population of zebra mussel (*Dreissena polymorpha* pallas). *Chemosphere*, 38, 3339-3356.

Gustavson, K. & Jonsson, P. (1999). Some Halogenated organic compounds in sediments and blue mussel (*Mytilus edulis*) in Nordic seas. *Marine Pollution Bulletin*, 38, 723-736.

Halldórsson, H.P., De Pirro, M., Romano, C., Svavarsson, J. & Sarà G. (2008). Immediate biomarker responses to benzo[a]pyrene in polluted and unpolluted populations of the blue mussel (*Mytilus edulis* L.) at high-latitudes. *Environment International*, 34, 486-489.

Hawkins, A.J.S. & Bayne, B.L. (1992). Physiological interrelations, and the regulation of production. In: Gosling, E. (Ed), The mussel *Mytilus*: ecology, physiology, genetics and culture (p 171-222). Amsterdam: Elsevier.

Hedgecock, D. (1994). Does variance in reproductive success limit effective population sizes of marine organisms? In: Beaumont, A. (Ed), Genetics and evolution of aquatic organisms (p 122-134). London: Chapman and Hall.

Hendriks, A.J., Pieters, H. & de Boer, J. (1998). Accumulation of metals, polycyclic (halogenated) aromatic hydrocarbons, and biocides in zebra mussel and eel from the Rhine and Meuse rivers. *Environmental Toxicology and Chemistry,* 17, 1885–1898.

Holt, T.J., Rees, E.I., Hawkins, S.J. & Seed, R. (1998). Biogenic Reefs (volume IX). An overview of dynamic and sensitivity characteristics for conservation management of marine SACs. Scottish Association for Marine Science (UK Marine SACs Project). 170 pp.

Humphreys, E. H., Janssen, S., Ilcil, A., Hiatt, P., Solomon, G. & Miller, M. D. (2008) Outcomes of the California ban on pharmaceutical lindane: Clinical and ecologic impacts. *Environmental Health Perspectives*, 116, 297-302.

Hunt, C.D. & Slone, E. (2010). Long-term monitoring using resident and caged mussels in Boston Harbor yield similar spatial and temporal trends in chemical contamination. *Marine Environmental Research*, 70, 343-357.

IEH a. 1995. Environmental estrogens: consequences to human health and wildlife. *Institute for Environment and Health*. Medical Research Council.

Igawa Martinez, D. & Fernandes Cardoso de Oliveira, J. (2010). Faecal bacteria in *Perna perna* (Linnaeus, 1758) (Mollusca:Bivalvia) for biomonitoring coastal waters and seafood quality. *Brazilian Journal of Oceanography*, 58, 29-35.

Isobe, T., Takada, H., Kanai, M., Tsutsumi, S., Isobe, K.O., Boonyatumanond, R. & Zalaroa, M.P. (2007). Distribution of Polycyclic Aromatic Hydrocarbons (PAHs) and phenolic endocrine disrupting chemicals in south and southeast Asian mussels. *Environmental Monitoring and Assessment*, 135, 423-440.

Jobling, S., Nolan, M., Tyler, C. R., Brighty, G. & Sumpter, J. P. (1998) Widespread sexual disruption in wild fish. *Environmental Science & Technology,* 32, 2498-2506.

Johns, C. & Timmerman, B.E. (1998). Total cadmium, copper and zinc in two dreissenid mussels, *Dreissena polymorpha* and *Dreissena bugensis*, at the outflow of Lake Ontario. *Journal of Great Lakes Research*, 24, 55–64.

Johnson, K., Wiederholm, T. & Rosenberg, D.M. (1993). Freshwater biomonitoring using individual organisms, populations and species assemblages of benthic macroinvertebrates. In: Rosenberg, D.M. & Resh, V. H. (Eds), Freshwater Biomonitoring and Benthic Macroinvertebrates (p 62-69). New York: Chapman & Hall.

Jonsson, H., Schiedek, D., Grøsvik, B. E. & Goksøyr, A. (2006). Protein responses in blue mussels (*Mytilus edulis*) exposed to organic pollutants: A combined CYP-antibody/proteomic approach. *Aquatic Toxicology*, 78, S49-S56

Jukes, T.H. & Shaffer, C.B. (1960). Antithyroid effects of aminotriazole. *Science*, 132, 296-297.

Julshamn, K. & Grahl-Nielsen, O. (1996). Distribution of trace elements from industrial discharges in the Hardangerfjord, Norway: A multivariate data analysis of Saithe, Flounder and Blue Mussel as sentinel organisms. *Marine Pollution Bulletin*. 32, 564–571.

Kauss, P.B. & Hamdy, Y.S. (1991). Polycyclic aromatic hydrocarbons in surficial sediments and caged mussels of the St. Marys River, 1985. *Hydrobiologia*, 219, 37–62.

Kramer, K.J.M., Jenner, H.A. & de Zwart, D. (1989) The valve movement response of mussels: a tool in biological monitoring. In: Munawar, A., Dixon, G., Mayfield, C.I., Reynoldson, T., & Sadar MH (Eds.). Environmental bioassay techniques and their application. (Hydrobiologia 188/189: p 433-443). Belgium: Kluwer Academic Publisher).

Köprücü, K., Yonar, S.M. & Şeker, E. (2010). Effects of cypermethrin on antioxidant status, oxidative stress biomarkers, behavior, and mortality in the freshwater mussel *Unio elongatulus eucirrus*. *Fisheries Science*, 76, 1007-1013.

Kwon, Y.K., Jung, Y.S., Park, J.C., Seo, J., Choi, M.S. & Hwang, G.S. (2012). Characterizing the effect of heavy metal contamination on marine mussels using metabolomics. *Marine Pollution Bulletin*, 64, 1874-1879.

Landis, W.G. & Yu, M.H. (1995). Introduction to environmental toxicology: Impacts of chemicals upon ecological systems. CRC Press, Inc.

Lauenstein, G.G., Robertson, A. & O'Connor, T.P. (1990). Comparison of trace-metal data in mussels and oysters from a mussel watch program of the 1970s with those from a 1980s program. *Marine Pollution Bulletin*, 21, 440-447.

Lintelmann, J., Katayama, A., Kurihara, N., Shore, L. & Wenzel, A. (2003). Endocrine disruptors in the environment - (IUPAC Technical Report). *Pure and Applied Chemistry,* 75, 631-681.

Lionetto, M., Caricato, R., Giordano, M. & Schetino, T. (2004). Biomarker Application for the Study of Chemical Contamination Risk on Marine Organisms in the Taranto Marine Coastal Area. *Chemistry & Ecology*, 20, 333-343.

Liu, J.H. & Kueh, C.S. (2005). Biomonitoring of heavy metals and trace organics using the intertidal mussel Perna viridis in Hong Kong coastal waters. *Marine Pollution Bulletin*, 51, 857-875.

Manley, A.R. & Davenport, J. (1979) Behavioral responses of some marine bivalves to heightened seawater copper concentrations. *Bulletin of Environmental Contamination and Toxicology*, 22, 739-744.

Marigómez, I., Soto, M., Cancio, I., Orbea, A., Garmendia, L. & Cajaraville, M.P. (2006). Cell and tissue biomarkers in mussel, and histopathology in hake and anchovy from Bay of Biscay after the Prestige oil spill (Monitoring Campaign 2003). *Marine Pollution Bulletin*, 53, 287-304.

Mersch, J., Jeanjean, A., Spor, H. & Pihan, J.C., (1992). The freshwater mussel *Dreissena polymorpha* as a bioindicator for trace metals, organochlorines and radionuclides. In: Neumann, D.,Jenner, H.A. (Eds). Zebra mussel (*Dreissena polymorpha*), Limnologie aktuell. (p 147–154). Stuttgart, Jena, New York: Gustav Fisher Verlag.

Metcalfe, J.L. & Charlton, M.N. (1990). Freshwater mussels as biomonitors for organic industrial contaminants and pesticides in the St. Lawrence River. *Science of the Total Environment,* 97-98, 595-615.

Moles, A. & Hale, N. (2003). Use of physiological responses in *Mytilus trossulus* as integrative bioindicators of sewage pollution. *Marine Pollution Bulletin*, 46, 954-958.

Mora, P., Michel, X. & Narbonne, J.F. (1999). Cholinesterase activity as potential biomarker in two bivalves. *Environmental Toxicology and Pharmacology*, 7, 253-260.

Nakata, H., Shinohara, R., Nakazawa, Y., Isobe, T., Sudaryanto, A., Subramanian, A., Tanabe, S., Zakaria, M.P., Zheng, G.J., Lam, P.K., Kim, E.Y., Min, B.Y., We, S.U., Viet, P.H., Tana, T.S., Prudente, M., Frank, D., Lauenstein, G. & Kannan, K. (2012). Asia-Pacific mussel watch for emerging pollutants: Distribution of synthetic musks and benzotriazole UV stabilizers in Asian and US coastal waters. *Marine Pollution Bulletin*, 10, 2211-2218.

Narbonne, J.F., Daubèze, M., Baumard, P., Budzinski, H., Clérandeau, C., Akcha, F., Mora, P. & Garrigues, P. Biochemical markers in mussel, *Mytilus* sp., and pollution monitoring in European coasts: Data Analysis. (2001). In: Garrigues , H. Walker, C.H. & Narbonne, J.F. (Eds). Biomarkers in Marine Organisms (p 238-239). London: Elsevier Science.

O'Connor, T.P. (1996). Trends in chemical concentrations in mussel and oysters collected along the US coast from 1986– 1993. *Marine Environmental Research*, 41, 183–200.

Ortiz-Zarragoitia, M. & Cajaraville, M.P. (2006). Biomarkers of exposure and reproduction-related effects in mussels exponed to endocrine disruptors. *Archives of Environmental Contamination and Toxicology*, 50, 361-369.

Pal, A., Gin, K.Y., Lin, A.Y. & Reinhard, M.(2010). Impacts of emerging organic contaminants on freshwater resources: review of recent occurrences, sources, fate and effects. *Science of the Total Environment*, 408, 6062-6069.

Pasantes, J.J., Martínez-Expósito, M.J., Torreiro, A. & Méndez, J. (1996). The sister chromatid exchange test as an indicador of marine pollution: some factors affecting SCE frequencies in *Mytilus galloprovincialis*, *Marine Ecology Progress Series*, 143, 113-119.

Petrović, S., Ozretić, B., Krajnović-Ozretić, M, & Bobinac, D. (2001). Lysosomal MembraneStability and Metallothioneins in Digestive Gland of Mussels (*Mytilus galloprovincialis* Lam.) as Biomarkers in a Field Study. *Marine Pollution Bulletin*, 42,1373-1378.

Phillips, D.J.H. (1978). Common mussel *Mytilus edulis* as an indicator of trace-metals in Scandinavian waters. II. Lead, iron and manganese. *Marine Biology*, 46, 147-156.

Phillips, P. T. & Rainbow, P. S. (1993). Biomonitoring of Trace Aquatic Contaminants, London: Elsevier Science.

Pineiro, M.E.A., Lozano, J.S. & Yusty, M.A.L. (1995). Organochlorine compounds in mussels of the estuarine bays of galicia (north-west Spain). *Marine Pollution Bulletin*, 30, 484–487.

Pipe, R.K., Coles, J.A., Thomas, M.E., Fossato, V.U. & Pulsford, A.L. (1995). Evidence for environmentally derived immunomodulation in mussels from the Venice Lagoon. *Aquatic Toxicology*, 32,59–73.

Porte, C., Biosca, X., Solé, M. & Albaigés, J. (2001). The integrated use of chemical analysis, cytochrome P450 and stress proteins in mussels to assess pollution along the Galician coast (NW Spain). *Environmental Polltuion,* 112-261-268.

Rainbow, P.S. (2006). Biomonitoring of trace metals in estuarine and marine environments. Australasian Journal of Ecotoxicology, 2006, 107-122.

Rainbow, P.S. & Phillips, D.J.H. (1993). Cosmopolitan biomonitors of trace metals. *Marine Pollution Bulletin*. 26, 593-601.

Reeders, H. H., de Vaate, A. B. & Noordhuis, R. (1993). Potential of the zebra mussel (*Dreissena polymorpha*) for water quality management. In: Napel, T.F. & Schloesser, D. W. (Eds), Zebra Mussels: Biology, Impact and Control (pp. 439-451) Boca Raton, FL: Lewis.

Regoli, F. (1992). Lysosomal responses as a sensitive stress index in biomonitoring heavy metal pollution. *Marine Ecology Progress Series*, 84, 63–69.

Resh, V.H. (2007). Multinational, freshwater biomonitoring programs in the developing world: lessons learned from African and southeast Asian river surveys. *Environmental Management*, 39, 737-748.

Resh, V.H., & Giap, D. H. (2010). Biomonitoring Methods for the Lower Mekong Basin. Mekong River Commission, Vientiane, Lao P.D.R.

Richardson, B.J., Garnham, J.S. & Fabris, J.G. (1994). Trace-metal concentrations in mussels (*Mytilus edulis planulatus L*) transplanted into southern Australian waters. *Marine Pollution Bulletin*, 28, 392-396.

Richman, L. & Somers. K. (2005). Can we use zebra and quagga mussels for biomonitoring contaminants in the Niagara River? *Water Air and Soil Pollution*. 167, 155-178.

Richman, L. & Somers, K. (2010). Monitoring metal and organic contaminant trends through time using quagga mussels (*Dreissena bugensis*) collected from the Niagara River. *Journal of Great Lakes Research*, 36, 28–36.

Rickwood, C.J. & Galloway, T.S. (2004). Acetylcholinesterase inhibition as a biomarker of adverse effect. A study of *Mytilus edulis* exposed to thepriority pollutant chlorfenvinphos. *Aquatic Toxicology*, 67, 45-56.

Riget, F., Johansen, P., & Asmund, G. (1996). Influence of length on element concentrations in blue mussels (*Mytilus edulis*). *Marine Pollution Bulletin*, 32, 745–751.

Roberts, D. (1975). Effect of pesticides on byssus formation in the common mussel, *Mytilus edulis*. *Environmental Pollution*, 8, 241-253.

Robson, A. A., Garcia de Leaniz, C., Wilson, R.P. & Halsey, L.G. (2010). Behavioural adaptations of mussels to varying levels of food availability and predation risk. Journal of Molluscan Studies. 76, 348-353.

Roper, J.M., Simmers, J.W. & Cherry, D.S. (2001). Bioaccumulation of butyltins in *Dreissena polymorpha* at a confined placement facility in Buffalo, New York. *Environmental Pollution*, 111, 447-452.

Roslev, P., Iversen, L., Sønderbo, H., Iversen, N. & Bastholm, S. (2009). Uptake and persistence of human associated *Enterococcus* in the mussel *Mytilus edulis:* relevance for faecal pollution source tracking. *Journal of Applied Microbiology*, 107, 944–953.

Ruessler, D.S., Kernaghan, N.J., Wieser, C.M., Wiebe, J.J. & Gross, T.S. (2011). An assessment of potential contaminant effects on freshwater mussels in the South Florida Ecosystem. In Southeast Ecologica Science Centre U.S. Geological Survey, USA. http://fl.biology.usgs.gov/posters/Ecotoxicology/Contaminants_Effect_Mussels/contaminants_effect_mussels.html.

Ruiz, J.M., Barreiro, R. & Gonzalez, J.J. (2005). Biomonitoring organotin pollution with gastropods and mussels. *Marine Ecology Progress Series*, 287, 169-176

Sabik, H., Gagne, F., Blaise, C., Marcogliese, D.J. & Jeannot, R. (2003). Occurrence of alkylphenol polyethoxylates in the St. Lawrence River and their bioconcentration by mussels (*Elliptio complanata*). *Chemosphere*, 51, 349-356.

Sáenz L.A., Seibert, E.L., Zanette, J., Fiedler, H.D., Curtius, A.J., Ferreira, J.F., Alves de Almeida, E., Marques, M.R. & Bainy, A.C.(2010). Biochemical biomarkers and metals in *Perna perna* mussels from mariculture zones in Santa Catarina, Brazil. *Ecotoxicology and Envrionmental Safety*, 73, 796-804.

Sálanki, J. & Varanka, L. (1978). Effect of some insecticides on the periodic activity of the fresh-water mussel (*Anodonta cygnea L.*). *Acta Biologica Academiae Scientarum Hungaricae*, 29, 173-180.

Scarpato, R., Migliore, L., Alfinito-Cognetti & Barale, R. (1990). Induction of micronuclei in gill tissue of *Mytilus galloprovincialis* exposed to polluted marine waters. *Marine Pollution Bulletin*, 2, 74-80.

Seed, R. & Suchanek, T.H. (1992). Population and community ecology of Mytilus. In: Gosling E.M. (Ed), The mussel *Mytilus*: ecology, physiology, genetics and culture (pp. 87-169). Amsterdam: Elsevier Science Publisher.

Selegean, J.P.W., Kusserow, R., Patel, R., Heidtke, T.M. & Ram J.L. (2001). Using zebra mussels to monitor *Escherichia coli* in environmental waters. *Journal of Environmental Quality*, 30, 171-179.

Serafim, A., Lopes, B., Company, R., Cravo, A., Gomes, T., Sousa, V. & Bebianno, M.J. (2011). A multi-biomarker approach in cross-transplanted mussels *Mytilus gallprovincialis*. *Ecotoxicology*, 20, 1959-1974.

Sinton, L.W., Hall, C.H., Lynch, P.A. & Daviescolley, R.J. (2002). Sunlight inactivation of fecal indicator bacteria and bacteriophages from waste stabilization pond effluent in fresh and saline waters. Applied and Environmental Microbiology, 68, 1122-1131Smolders, R. & Schoeters, G. (2007). Identifying opportunities and gaps for establishing an integrated EDR-triad at a European level. *International Journal of Hygiene and Environmental Health,* 2010, 253-257.

Slooff, W., Zwart, D. & Marquenie, J.M. (1983). Detection limits of a biological monitoring system for chemical water pollution based on mussel activitiy. *Bulletin of Environmental Contamination and Toxicology*, 30, 400-405.

Solé, M., Porte, C., Barcelo, D., & Albaiges, J. (2000). Bivalves residue analysis for the assessment of coastal pollution in the Ebro Delta (NW Mediterranean). *Marine Pollution Bulletin*, 40, 746-753.

Stäb, J.A., Frenay, M., Freriks, I.L., Xofino, W.P. & Brinkman, U.A.T. (1995). Survey of nine organotin compounds in the netherlands using the zebra mussel (*Dreissena Polymorpha*) as biomonitor. *Environmental Toxicology and Chemistry,* 14, 2023-2032.

Sudaryanto, A., Takahashi, S., Monirith, I., Ismail, A., Muchtar, M., Zheng, J., Richardson, B.J., Subramanian, A., Prudente, M., Hue, N.D. & Tanabe, S. Asia-Pacific mussel watch: monitoring of butyltin contamination in coastal waters of Asian developing countries. *Environmental Toxicology and Chemistry*, 21, 2119-2130.

Sunlu, U. (2006). Trace metal levels in mussels (*Mytilus galloprovincialis* L. 1758) from Turkish Aegean Sea coast. *Environmental Monitoring and Assessment,* 114, 273–86.

Szefer, P., Fowler, S.W., Ikuta, K., Paez Osuna, F., Ali, A.A., Kim, B.S., Fernandes, H.M., Belzunce, M.J., Guterstam. B., Kunzendorf, H., Wolowicz, M., Hummel, H. & Deslous-Paoli M. (2006). A comparative assessment of heavy metal accumulation in soft parts and byssus of mussels from subarctic, temperate, subtropical and tropical marine environments. *Environmental Pollution*, 139, 70-78.

Tanabe, S., Prudente, M.S., Kan-Atireklap, S. & Subramanian, A. (2000). Mussel watch: marine pollution monitoring of butyltins and organochlorines in coastal waters of Thailand, *Philippines and India.Ocean and Coastal Management*, 43, 819–839.

Tavares, T.M., Beretta, M. & Costa, M.C., (1999). Ratio of DDT/DDE in the All Saints Bay, Brasil and its use in environmental management. *Chemosphere*, 38, 1445–1452.

Thébault, H., Arnaud, M., Charmasson, S., Andral, B., Diméglio, Y. & Barker, E. (2005). Bioavailability of anthropogenic radionuclides in mussels along the French Mediterranea coast. *Radioprotection*, 40, S47-S52.

Thompson, S., Budzinski, H., Garrigues, P. & Narbonne, J.F., (1999). Comparison of PCB and DDT distribution between water-column and sediment-dwelling bivalves in Arcachon Bay, France. *Marine Pollution Bulletin*, 38, 655–662.

Usero, J., Gonzalez-Regalado, E. & Gracia, I. (1996). Trace metals in the bivalve mollusc *Chamelea gallina* from the Atlantic Coast of Southern Spain. *Marine Pollution Bulleting*, 32, 305-310.

Valette-Silver, N.J. & Lauenstein, G.G. (1995).Radionuclide concentrations in bivalves collected along the coastal United States. *Marine Pollution Bulletin*, 30,320-331.

van der Oost, R., Beyer, J. & Vermeulen, N.P.E. (2003). Fish bioaccumulation and biomarkers in environmental risk assessment: a review. *Environmental Toxicology and Pharmacology*, 13, 57-149.

Venier, P., Pallavicini, A., De Nardi, B. & Lanfranchi, G. (2003). Towards a catalogue of genes transcribed in multiple tissues of *Mytilus galloprovincialis. Gene*, 314,29-40.

Viarengo, A., Lowe, D., Bolognesi, C., Fabbri, E. & Koehler, A. (2007). The use of biomarkers in biomonitoring: a 2-tier approach assessing the level of pollutant-induced stress syndrome in sentinel organisms. Comparative Biochemistry and Physiology, Part C: *Toxicology & Pharmacology*, 146, 281-300.

Villela, I.V., de Oliveira, I.M., Silveira, J.C., Dias, J.F., Henriques, J.A. & da Silva, J. (2007). Mutational Research. Assessment of environmental stress by the micronucleus and comet assays on Limnoperna fortunei exposed to Guaíba hydrographic region samples (Brazil) under laboratory conditions, 628, 76-86.

Vukmirović, M., Bihari, N., Zahn, R.K.. & Müller, W.E.G. & Batel, R. (1994). DNA damage in marine mussel *Mytilus galloprovincialis* as a biomarker of environmental contamination. *Marine Ecology Progress Series*, 109, 165-171.

Wolf, M.S. & Toniolo, P.G. (1995). Environmental organochlorine exposure as a potential etiologic factor in breast cancer. Envrionmental Health Perspectives, 103, 141–145.Yap, C.K., Ismail, A., Cheng, W.H. & Tan S.G. (2006). Crystalline style and tissue redistribution in Perna viridisas indicators of Cu and Pb bioavailabilities and contamination in coastal waters. *Ecotoxicology and Environmental Safety*, 63, 413-423.

Wu, R.S.S., Lau, T.C., Fung, W.K.M., Ko, P.H. & Leung, K.M.Y. (2007). An "artificial" mussel for monitoring heavy metals in marine environments. *Environmental Pollution*, 1451 104-110.

Yaqin, K., Lay, B.W., Riani, E., Masud. S.A. & Hansen, P.D. (2011). Hot spot biomonitoring of marine pollution effects using cholinergic and immunity biomarkers of tropical green mussel (*Perna viridis*) of the Indonesian waters. *Journal of Toxicology and Environmental Health Sciences*, 3, 328-344.

Zippay, M. & Helmut, B. (2012). Effects of temperature change on mussel, *Mytlius*. *Integrative Zoology*, 7, 312-327.

In: Mussels: Ecology, Life Habits and Control
Editors: Jarek Nowak and Michal Kozlowski
ISBN: 978-1-62618-083-3

Chapter 4

BIOENERGETIC MODEL PREDICTIONS OF ACTUAL GROWTH AND ALLOMETRIC TRANSITIONS DURING ONTOGENY OF JUVENILE BLUE MUSSELS *MYTILUS EDULIS*

***Poul S. Larsen*[1], *Kim Lundgreen*[2] *and Hans Ulrik Riisgård*[2*]**
[1]DTU Mechanical Engineering, Fluid Mechanics, Technical University of Denmark, Kgs. Lyngby, Denmark
[2]Marine Biological Research Centre (University of Southern Denmark), Kerteminde, Denmark

ABSTRACT

The growth rates of blue mussels *Mytilus edulis* on ropes in the Great Belt (Denmark) have been studied during the growth season of one year, from settling to about 30 mm shell length mussels, covering > 4 decades of body mass. Measured shell length (L, mm) and dry weight of soft parts (W, µg) for $L > 10$ mm followed a power-law ($W = 2.15L^{3.40}$) which supplemented an existing power-law for $L < 10$ mm ($W = 24.7\,L^{2.42}$) to establish that somatic growth changes character at $L \approx 10$ mm and $W \approx 10$ mg. Results of specific growth rates based on dry weight of soft parts ($\mu = \mathrm{dln}W/\mathrm{dt}$) compared well with predictions based on a previously developed bioenergetic growth model (BEG) for $W > 10$ mg ($\mu = aW^b$, $a = 0.871\times C - 0.986$; $b = -0.34$, with μ in % d^{-1} and W in g) which explicitly takes into account the prevailing chl *a* concentration C (µg L^{-1}). Results for $W < 10$ mg also correlated well by the power-law ($\mu = aW^b$), now with exponent $b = -0.13$ close to the suggested value ($b \approx -0.1$) from experimentally established correlations for filtration and respiration rate of post-metamorphic mussels. Using the stated $W(L)$-relation for $L > 10$ mm the growth model has been expressed in terms of shell length specific growth rate ($\mu_L \equiv \mathrm{dln}L/\mathrm{dt} = \alpha L^{\beta}$) by which data on shell length was well correlated, including the influence of chl *a* concentration. Supplementary growth data from mussels in suspended net-bags at the same site illustrated differences ascribed to lack of competition for space and food, and literature data on shell length from cage-

[1] E-mail: psl@mek.dtu.dk.
[*] Corresponding author, E-mail: hur@biology.sdu.dk.

growth of mussels in the brackish Baltic Sea support the present correlations. It is argued that the allometric transitions that take place around $W \approx 10$ mg and $L \approx 10$ mm during the ontogeny of *M. edulis* is most likely universal and not restricted to first year growth of juvenile (young) mussels during the productive season.

Keywords: Energy budget, specific growth rate, ontogeny, *Mytilus edulis*, allometric transitions

INTRODUCTION

The common blue mussel, *Mytilus edulis* L. may reproduce at any time of the year, but in the temperate zone spawning is most frequent during spring and summer (Seed 1976; Dare 1976). Temperature seems to be the most important factor that determines the annual reproductive cycle, and a cold winter tends to synchronize spawning the following spring (Savage 1956; Jørgensen 1981). The fertilized egg develops to a trochophora larva that soon after becomes a free swimming veliger larva feeding on small 2 to 4 µm particles (Bayne 1965; Riisgård et al. 1980; Jespersen and Olsen 1982; Sprung 1984; Gosling 2003). This stage lasts between 3 and 5 weeks (Thorson 1961), i.e. time to reach ~300 µm, depending on environmental factors (temperature, food ration, salinity) until metamorphosis when an extensible foot appears and the larva becomes a pediveliger that settles to seek out a suitable substrate. After attachment, morphological change (metamorphosis) results in a young post-metamorphic mussel (Bayne 1965; Dare 1976; Gosling 2003; Dolmer and Stenalt 2010). The subsequent growth has been extensively investigated in both field studies, as a prerequisite to assessment of the growth potential and its exploitation in nature, and in laboratory studies (Jørgensen 1976, 1990; Stirling and Okumus 1994; Okumus and Stirling 1994; Clausen and Riisgård 1996). Many studies of growth in nature have been based on measurements of increases in shell length (e.g. Theisen 1975; Kautsky 1982; Kristensen and Lassen 1997; Dolmer 1998; Westerbom et al. 2002; Vuorinen et al. 2002) and it is therefore of interest to know the relations between growth in shell length and weight of soft parts. The present study aims at showing how bioenergetic model considerations may guide the data analysis of young post-metamorphic and juvenile/adult *M. edulis* during the productive period of a year.

The growth of filter-feeding mussels is primarily dependent on the amount of food ingested, which is closely related to the filtration rate, and the energy used in metabolism, frequently measured as the oxygen uptake rate or respiration rate. The filtration rate (F, L h^{-1}) of *Mytilus edulis* can be estimated from the dry weight of soft parts (W, g) according to the formula (Møhlenberg and Riisgård 1979): $F = a_1W^{b1}$, where $a_1 = 7.45$ and $b_1 = 0.66$, and further, the respiration rate (R, µL O_2 h^{-1}) can be estimated according to the formula (Hamburger et al. 1983): $R = a_2W^{b2}$, where $a_2 = 475$ and $b_2 = 0.663$. Because $b_1 \approx b_2 = 0.66$ for mussels > 10 mg tissue dry weight, it has recently been suggested by Riisgård et al. (2012a) that the weight specific growth rate may be expressed as: $\mu = aW^b$, where $b = b_1 - 1 = -0.34$. This bioenergetic growth model (BEG) predicts that the specific growth rate of mussels > 10 mg is a power function with $b = -0.34$. However, for smaller post-metamorphic mussels <10 mg dry tissue weight Hamburger et al. (1983) found that the respiration rate (R, µL O_2 h^{-1}) could be expressed as: $R = 315W(g)^{0.887}$. Further, for post-metamorphic *M. edulis* (total organic dry weight, $W = 70$ to 10000 µg) Riisgård et al. (1980) found that the filtration rate (F, mL h^{-1}) could be expressed as: $F = 0.025W(\mu g)^{1.03}$. Thus for young post-metamorphic mussels < 10 mg, $b_1 \approx b_2 \approx 0.9$ which implies that the BEG model predicts $b \approx -0.1$. In a recent study on juvenile *M. edulis* Riisgård et al. (2012b) found that the ingestion rate ($I = F \times C$) increased linearly up to the incipient saturation concentration (C_{sat}; see also Riisgård et al. 2011a) and remained nearly constant above C_{sat}. But the measured weight specific growth rates (μ) decreased sharply above C_{sat} from a maximal value of about 9.5 % d^{-1} to about 1.5% d^{-1}. Below C_{sat} on the other hand, measured μ-values increased linearly with increasing algal concentration in agreement with the BEG model. Thus, the response to increasing food concentration with possible regulation of net ingestion may only come into play when C_{sat} is exceeded (Riisgård et al. 2012b).

The above considerations suggest it appropriate to present field data on mussel growth as weight specific growth rate versus dry weight of soft parts and correlate results by power-laws as done by Riisgård et al. (2012a). Among other models is the frequently used SFG (scope for growth) that is based on the energy balance, where energy available for growth is calculated from the difference between assimilated energy and respiration (e.g. Bayne and Widdows 1978; Bayne and Worrall 1980; Hawkins et al. 2001). The different energetic parameters have typically been measured in the laboratory on mussels from different sites, and the physiological response (i.e. 'scope for growth') has been used as a measure of pollution stress (e.g. Widdows et al. 1995). Two other models, MTE (metabolic theory of ecology) (Brown et al. 2004; van der Meer 2006), and DEB (dynamic energy budget) (e.g. Beadman et al. 2002; van der Veer et al. 2006; Bourlès et al. 2009; Rosland et al. 2009; Duarte et al. 2010; Saraiva et al. 2011; Filgueira et al. 2011) are likewise based on the energy budget. Here, the latter is the most general model, considering the three levels of: ecosystem, population and individual where our BEG model falls at the individual level as a "physiological component model" and has a convenient form that suggests a simple way to correlate actual field data on growth, foremost because ingestion and maintenance rates depend on the approximately same powers of size. Thus, using the simplest forms (van der Meer 2006, [Eq.1] or [Eq.4], respectively, therein), the MTE or DEB model implies that ingestion and maintenance rates depend on dry weight of soft parts (W) to powers of $b_1 = 3/4$ and $b_2 = 1$ or $b_1 = 2/3$ and $b_2 = 1$, respectively, which leads to the forms $\mu = c_1W^{-1/4} - c_2$ or $\mu = c_3W^{-1/3} - c_4$, respectively, where coefficients c_i depend on environmental parameters. Each of the stated forms implies growth to a finite maximal size but neither form is convenient to correlate experimental data during the productive growth period which excludes the terminal stage approaching maximal size. Also, the exponent of $b_1 = 3/4$ for ingestion is not universally valid for the whole ontogeny but has been found to be $b_1 = 0.66$ and $b_1 \approx 1$, respectively for larger and smaller blue mussels (Riisgård et al. 1980; Hamburger et al. 1983; Riisgård 1998). It should also be noted that the prediction of growth in time, given environmental parameters as pursued in MTE- and DEB-models, is a far more challenging task than the prediction of rate of growth of a given size class as pursued in the BEG-model. The present study of juvenile mussels examines the maximum individual growth rate of mussels on suspended ropes in the Great Belt (Denmark) from the early juvenile stage to young adult, and the field data are compared to bioenergetic model predictions. Additional data on mussels growing in suspended net-bags at the same site serve as reference for the growth potential when possible intra-specific competition for space and/or food on mussel-ropes is eliminated.

Materials and Methods

Mussel-Ropes

Mussel-rope bands (5 m long by 5 cm wide, hereafter 'mussel-ropes') were set out in May 2011 at the MarBioShell research and demonstration mussel farm in Kerteminde Bay, Denmark (Figure 1), and subsequently collected at different intervals between 14 June and 9 December 2011. The densest part of the top end (1.5 m) of the farm-rope was located and a suitable piece of the mussel-rope was cut off for determination of shell length of settled mussels. Mussels on the rope were scraped off and counted along with the individuals still attached to the farm-rope by using a stereo microscope. Subsequently, shell length was determined for 3 groups; 1: a random sample of all mussels, 2: a sample of the 20 % biggest mussels, and 3: a sample of the subjectively selected 10 biggest individuals on each mussel-rope. For the samples collected later in the growth season (when mussels were >10 mm) dry weight of soft parts was also determined for Group 3 by drying the soft parts on pieces of tin foil in an oven for 24 h at 90 °C.

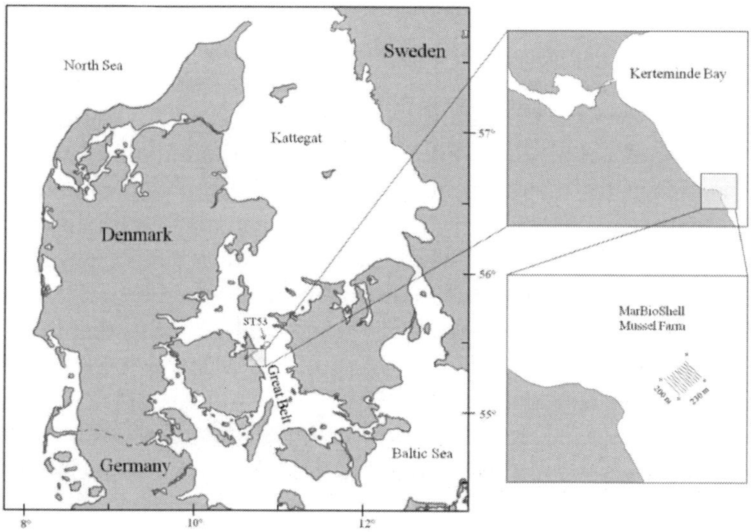

Figure 1. Map of Denmark showing the location of the MarBioShell research and demonstration mussel-farm with 10 lines for mussel production in the Great Belt (Kerteminde Bay). Farm-ropes, Series#1b and Series#2 net bags were placed on the most eastern line of the mussel farm whereas Series#1a net-bags were placed in the middle of the farm. Temperature, salinity and chl *a* were measured at ST53.

Net Bags

Mussels were collected from the MarBioShell mussel farm in September 2011 and transported to the nearby Marine Biological Research Centre (SDU) in Kerteminde and kept in large aquaria with running seawater prior to preparation for two series of field experiments with mussels hung up in net-bags.

Series #1. Mussels were divided into two size groups (mean shell length ±S.D.; 25.8±0.7 mm, n = 12 and 31.8±0.4 mm, n = 17) and put into net bags (Go Deep International Inc.). The net bags with mussels were transported to the field location site and hung up in a buoy system at the outer edge of the mussel farm 3 days before the experimental period started to allow the mussels to attach to the net material and acclimatize. At the start of the experimental period (4 October 2011; Day 0) some mussels of each of the two size groups were collected for analysis and some were transferred to the middle of the mussel farm (Figure 1). Subsequently, mussels of each size group at each location (#1a: middle and #1b: outer edge of the farm) were collected for analysis on Day 7 and Day 16. A record of chlorophyll *a* concentration, temperature and salinity near the mussel farm during the year 2011 were obtained from the Danish Nature Agency, Ministry of the Environment (Figure 2).

Series #2. Mussels were divided into two size groups (mean ±S.D.; 26.6±1.2 mm, n= 12 and 32.7±0.5, n = 17) and put into net bags (Go Deep International Inc.). The net bags with mussels were transported to the field location site and hung up in a buoy system at the outer edge of the mussel farm (Figure 1) 2 days before the experimental period started to allow the mussels to attach to the net material and acclimatize. At the start of the experimental period (10 November 2011; Day 0) mussels of each size group were collected. Subsequently, mussels of each size group were collected on Day 8 and Day 13. For both Series #1 and Series #2, collected mussels from net bags were analyzed for shell length and dry weight of soft parts.

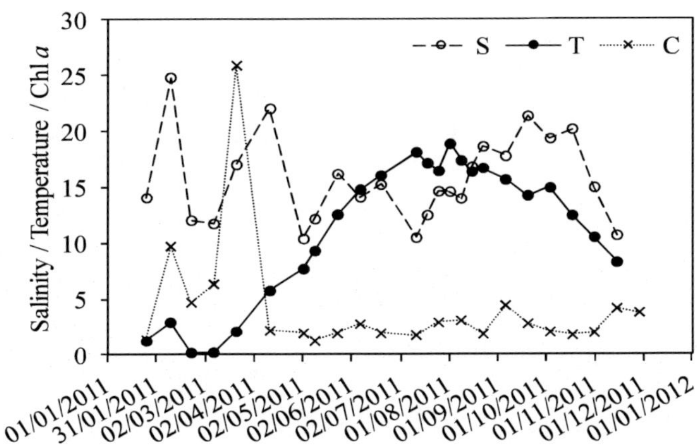

Figure 2. Environmental parameters in Great Belt during 2011. Mean (±S.D.) values were: 15.6±3.9, 11.2±6.2 and 4.1±5.2 for salinity (S, psu), temperature (T, °C) and chl *a* (C, μg L^{-1}). The high peak in chl *a* (25.9 μg L^{-1}) in March was caused by a bloom of *Chatonella* (Pseudochattonella farcimen). Mean (±S.D.) chl a concentration during 2011 after the bloom was 3.1±2.1 μg L^{-1}.Data obtained from Danish Nature Agency.

Equations

The dry weight of soft parts (W, μg) of young post-metamorphic *Mytilus edulis* (shell length < 10 mm) was calculated using (Jespersen and Olsen 1982):

$$W = 24.7L^{2.42}\,, \qquad (1)$$

where L is shell length in mm. The condition index (CI, mg cm^{-3}) is given by

$$CI = W/L^3 \,, \tag{2}$$

where the dry weight of larger mussels was determined directly (90 °C, 24 h). Weight specific growth rates of mussels (μ, % d^{-1}) was calculated from

$$\mu = \ln(W_t/W_0)/\Delta t \,, \tag{3}$$

(or equivalently from the slope of the trend line in a plot of ln$W$ versus time) where W_t and W_0 are dry weight of soft parts at the end and start of growth period, respectively, and Δt is length of period in days. The average value of μ obtained this way is taken to be valid at the average dry weight of soft parts defined by

$$W_{avg} = (W_0 \times W_t)^{1/2} \,. \tag{4}$$

The estimated weight specific growth rate (μ_{est}, % d^{-1}) for mussels > 10 mg according to the BEG model (Riisgård et al. 2012a, Eq.18 therein) is

$$\mu_{est} = \mathrm{a}W^{b}, \ (a = 0.871 \times C - 0.986; \ b = -0.34) \,, \tag{5}$$

where W (g) is dry weight of soft tissue and C (μg L^{-1}) the chlorophyll a concentration. Using the definition, $\mu = (1/W)\mathrm{d}W/\mathrm{d}t$, and the power-law of Eq.(5), i.e. $\mu \equiv (1/W)\mathrm{d}W/\mathrm{d}t = aW^b$, or $W^{-(1+b)}\mathrm{d}W = -(1/b)\mathrm{d}W^{-b} = a\mathrm{d}t$, the time history of growth from a given initial size (W_1 at time t_1) is obtained by integration, yielding $W^{-b} - W_1^{-b} = -ba(t-t_1)$, or

$$W = [W_1^{-b} + (-ba)(t - t_1)]^{-1/b} \,. \tag{6}$$

In the same way, defining the shell length specific growth rate (μ_L, d^{-1}) by

$$\mu_L \equiv (1/L)\mathrm{d}L/\mathrm{d}t = \mathrm{d}\ln L/\mathrm{d}t \approx \ln(L_t/L_0)/\Delta t \,, \tag{7}$$

where L_t and L_0 are shell length at the end and start of growth period, respectively, and Δt is length of period in days, and assuming the validity of a power-law for shell growth over some size range,

$$\mu_L = \alpha L^{\beta} \,. \tag{8}$$

The time history of shell growth from a given initial size (L_1 at time t_1) is obtained by integration (as for Eq.6), yielding

$$L = [L_1^{-\beta} + (-\beta\alpha)(t - t_1)]^{-1/\beta} \,. \tag{9}$$

Values of μ_L may be calculated from shell length data $L(t)$ in the same way as μ from $W(t)$ in Eq.(3), or equivalently from the slope of the trend line in a plot of ln$L$ versus time.

Average values of μ_L obtained this way over a time interval from 0 to t is taken to be valid at the average shell length defined by

$$L_{avg} = (L_0 \times L_t)^{1/2} . \qquad (10)$$

We may also relate μ_L to μ by introducing the power-law $W = cL^d$ into Eq.(5) which gives $\mu = \mathrm{dln}W/\mathrm{d}t = d\,\mathrm{dln}L/\mathrm{d}t = d\,\mu_L$, hence

$$\mu_L = \mu/d = (a/d)W^b , \qquad (11)$$

from which W is finally eliminated by using again $W = cL^d$, leaving

$$\mu_{L,est} = \alpha L^{\beta} ; \alpha = a/(dc^b) , \beta = bd . \qquad (12)$$

Statistics

Pearson Product Moment Correlation (PPMC) was carried out in SigmaPlot and used to measure the strength of the correlation between dry weight of soft parts and weight specific growth rate, and shell length and shell length specific growth rate. The hypothesis that slopes were significantly different from zero was accepted for $P < 0.05$. Analysis of Covariance (ANCOVA) was carried out in Systat and used for investigation if slopes of correlations between dry weight of soft parts and weight specific growth rate, and shell length and shell length specific growth rate were different between groups of mussels. Hypothesis of significant difference of slopes was accepted for $P > 0.05$.

RESULTS

Veliger Larvae and Settling

The time domain of the study covered most of the year 2011. Figure 3 shows the occurrence of veliger larvae in the vicinity of the suspended mussel-ropes, averaging 2 ind. L^{-1} during May to June, and 1 ind. L^{-1} during July to August whereupon the density fades out. At the start of farm-rope data collection on 14 June, settled larvae had an average length of 0.6 mm (Table 1, Group 1). The density of settled individuals then quickly increased to peak late in July after which the density declined, but the recorded mean shell lengths of all of the 3 groups continued to increase (Figure 4, Table 1).

Mussel-Ropes

Growth versus time in terms of shell length (L) is summarized in Table 1 and Figure 4 for the 3 selected groups of mussels: Group 1 (s) random samples of all sizes, Group 2 (20 %) the 20 % biggest individuals, and Group 3 (10) the 10 biggest individuals. Growth versus time in terms of dry weight of soft parts (W) is summarized in Table 2 for Group 3 of 10 biggest mussels, both as estimated from shell length according to Eq.(1) and, for $W > 10$ mg as measured directly. Using these results and Eq.(3) give the tabulated values of weight specific growth rates (μ) versus average dry weight (W_{avg}, Eq.4), also shown in the log-log plot of Figure 5. To show the change in weight specific growth rates versus time we have plotted $\ln W$ versus time in Figure 6 where the slope of lines through successive points, according to Eq.(3), gives the same average value μ over time intervals as listed in Table 2. Further, the exponent of a regression line through the first 3 of the 4 data points from measured dry weight of soft parts (W_0 of Table 2) versus time, Figure 7, gives the average specific growth rate ($\mu = 5.37\ \%\ d^{-1}$) during the period from 27 July to 4 October.

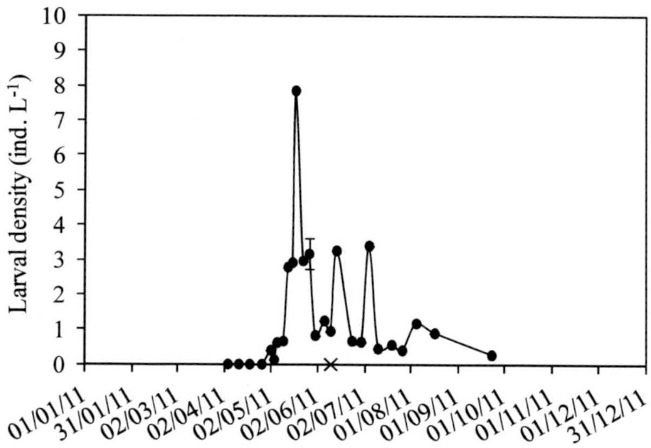

Figure 3. *Mytilus edulis*. Density of mussel veliger larvae at the experimental site. The cross marks the date (10 June 2011) when the first settling of pediveliger larvae on farm-ropes was observed.

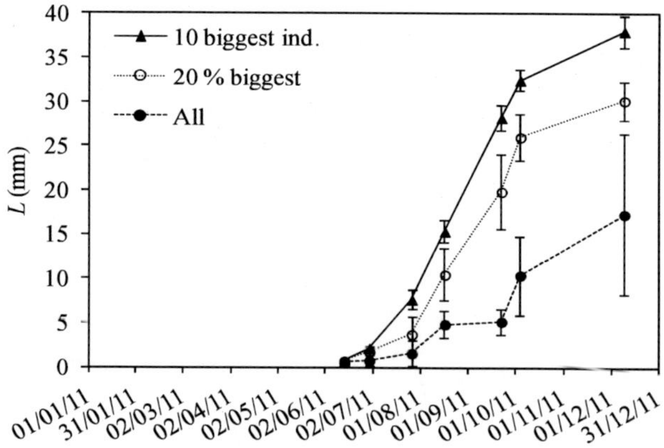

Figure 4. *Mytilus edulis* (farm-ropes, Groups 1, 2 and 3). Mean (±S.D.) shell length (L, mm) mussels for each of 3 groups: (1) all mussels on farm-ropes, (2) biggest 20%, and (3) 10 biggest individuals found on the rope.

Table 1. *Mytilus edulis* (farm-ropes).
Growth parameters for mussels in the period
14 June to 9 December 2011

		Rope-segment			Group 1			Group 2					Group 3			
Date 2011	Δt (d)	n_{tot}	RS (cm)	D (ind. cm^{-1})	n_s	L (mm)	ΔL (mm mo^{-1})	$n_{20\%}$	L (mm)	ΔL (mm mo^{-1})	L_{avg} (mm)	μ_L (d^{-1})	L (mm)	ΔL (mm mo^{-1})	L_{avg} (mm)	μ_L (d^{-1})
14-06	0	48	5	10	48	0.6±0.1		10	0.8±0.1				0.8±0.1			
30-06	16	292	5	58	146	0.8±0.6	0.5	29	1.8±0.3	2.0	1.17	0.0848	2.2±0.3	2.7	1.28	0.0632
27-07	27	1454	5	291	522	1.7±1.5	1.0	104	3.8±2.1	2.2	3.84	0.0379	7.7±1.1	6.2	4.10	0.0464
17-08	21	864	5	173	145	4.9±1.5	4.7	29	10.5±3.0	9.8	9.53	0.0379	15.4±1.3	11.2	10.90	0.0330
22-09	36	4453	21	212	255	5.2±1.4	0.3	51	19.9±4.2	7.9	19.18	0.0167	28.3±1.4	10.8	20.88	0.0169
11-10	12	3741	44	85	136	10.5±4.4	13.2	27	26.1±2.6	15.6	27.50	0.0100	32.5±1.2	10.8	30.31	0.0115
09-12	66	1855	35	53	91	17.4±9.1	3.2	18	30.3±2.2	1.9	30.23	0.0011	38.1±1.8	2.6	35.19	0.0024

Δt = days since last subsample; n_{tot} = total number of mussels on farm-rope segment sample; n_s = number of randomly collected mussels in sample for shell length measurement (Group 1); $n_{20\%}$ = number of mussels from the 20 % biggest individuals of the sample n_s (Group 2); 10 biggest mussels found on farm-rope (Group 3); RS = rope segment; D = density; L = shell length of mussels in group; ΔL = increase in shell length since last subsample; L_{avg} = average shell length, Eq.(10); μ_L = shell length specific growth rate, Eq.(7). Mean±S.D. are shown.

Table 2. ***Mytilus edulis*** **(farm-ropes, Group 3: 10 biggest individuals)**

				Calculated Eq. (1)				Measured			
Growth period 2011	Δt (d)	T (°C)	C (μg L^{-1})	W_0 (mg)	W_{end} (mg)	W_{avg} (mg)	μ_{cal} (% d^{-1})	W_0 (mg)	W_{end} (mg)	W_{avg} (mg)	μ (% d^{-1})
14 Jun – 30 Jun	16	16.0	1.9	0.013	0.163	0.05	16.0				
30 Jun – 27 Jul	27	17.2±0.8	2.3±0.8	0.163	3.45	0.75	11.3				
27 Jul – 17 Aug	21	17.5±1.3	3.1	3.45	18.6	8.0	8.0	3.45*	23.5	9.0	9.1
17 Aug – 22 Sep	36	15.5±1.2	3.0±1.3	18.6	80.2	38.6	4.1	23.5	182.3	65.5	5.7
22 Sep – 04 Oct	12	14.9	2.0	80.2	112.7	95.1	2.8	182.3	296.0	232.3	4.0
04 Oct – 09 Dec	66	11.5±2.8	2.9±1.2	112.7	165.3	136.5	0.6	296.0	242.3	267.8	-0.3

Growth periods, lengths of growth periods (Δt), mean temperature in period (T), mean chl *a* (C), dry weight of soft parts at start (W_0) and end (W_{end}) of growth periods, average dry weight of soft parts in growth period (W_{avg}, Eq.4) and corresponding weight specific growth rates (μ, Eq.3) are presented as calculated from shell length data using Eq. (1), and as actually measured body dry weight on collected mussels with shell length >10 mm. Mean±S.D. are shown.*Calculated.

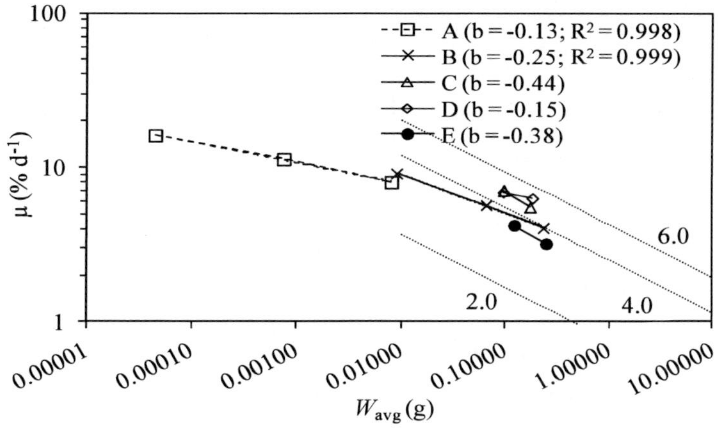

Figure 5. *Mytilus edulis*. Weight specific growth rate (μ, from Tables 2 and 3) versus average body dry weight (Eq.4). Power-law exponent of regression lines (*b*) are shown in the figure. A: small farm-rope mussels ($W_{avg} < 10$ mg), B: larger farm-rope mussels ($W_{avg} > 10$ mg), C and D: net-bag mussels, Series #1a, 1#b in October 2011, and E: Series #2 in November 2011. Dotted lines: growth model (Eq.5) for 3 values of chl *a* concentration (μg L^{-1}).

As a supplement to the allometry of Eq.(1) valid for young post-metamorphic *Mytilus edulis* (shell length < 10 mm) a plot (Figure 8) of the measured values of both dry weight of soft parts (*W*, μg) and shell length (*L*, mm) of larger mussels (> 10 mm) during growth (from 17 August to 4 October) shows good agreement with the power law: $W = 2.15L^{3.40}$. To show growth of shell length in terms of shell length specific growth rate (Eq.7) we have plotted ln*L* versus time (not shown but similar to Figure 6 for *W*) in which the slope of lines through successive points gives the average value μ_L over time intervals as listed in Table 1 for Groups 2 and 3. These results versus average length (L_{avg}, Eq.10) are shown in the log-log plot of Figure 9 in which coefficients (α and β of Eq.12) of regression lines are given.

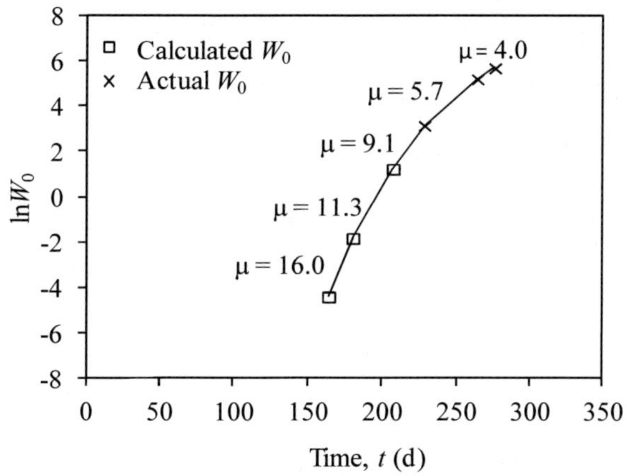

Figure 6. *Mytilus edulis* (farm-ropes, Group 3). Semi-logarithmic plot of dry weight of soft parts (W_0, mg; Table 2), either calculated from measured shell length (Eq.1) of small post-metamorphic mussels (□) or from actual measurement of soft parts (x), versus time. The weight specific growth rates (μ, % d^{-1}) calculated by means of Eq.(3) for all growth periods are indicated.

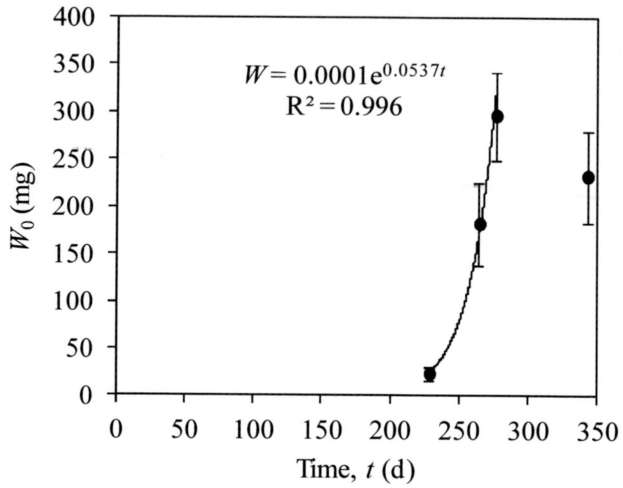

Figure 7. *Mytilus edulis* (farm-ropes, Group 3). Measured dry weight of soft parts (W_0, Table 2) of the 4 last samples versus time (Day 0 = 1 January 2011. The exponent of the exponential trend line, excluding the last sample, shows a mean weight specific growth rate (μ) of 5.37 % d^{-1}.

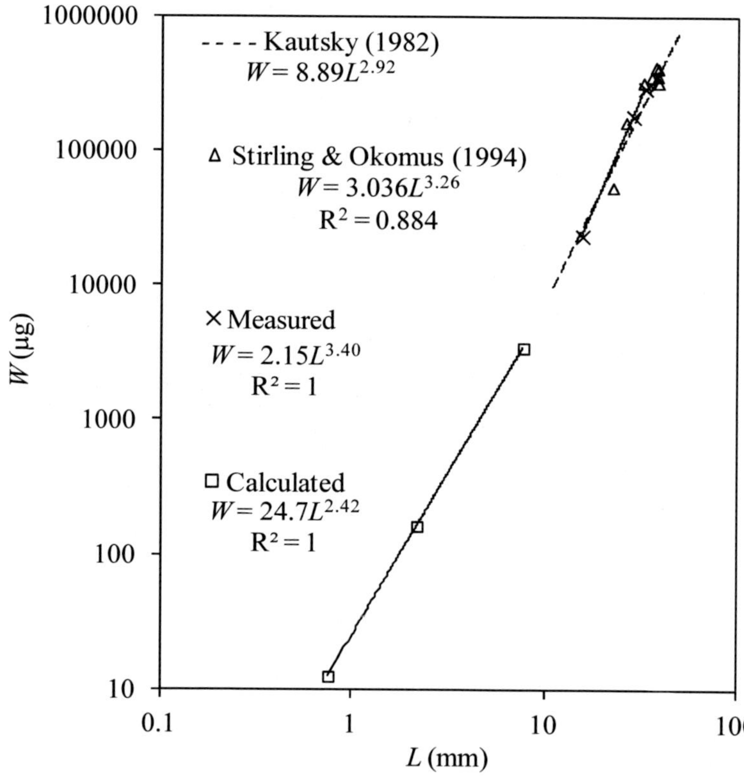

Figure 8. *Mytilus edulis* (farm-ropes, Group 3). Dry weight of soft parts calculated (Eq.1) for young post-metamorphic mussels (□) and measured (Eq.13) for larger juvenile mussels (x), both versus measured shell length (Table 2). Correlation of Kautsky (1982, Figure 17 therein) (dotted line). $W(L)$ from Stirling and Okumus (1994, their Series LL) includes data from May to November 1991 (Δ), after which chl a concentration falls below 0.2 μg L^{-1} and dry weight started to decrease.

Net-Bags

Growth in terms of shell length (L), measured dry weight of soft parts (W), calculated specific growth rate (μ) and condition index (CI) for the 2 size groups of mussels at each of Series #1a, #1b and Series #2 are summarized in Table 3 and Figures 10 and 11. The determined values of μ and μ_L from these series have also been shown in Figure 5 and Figure 9, respectively, for comparison with results from mussel-ropes.

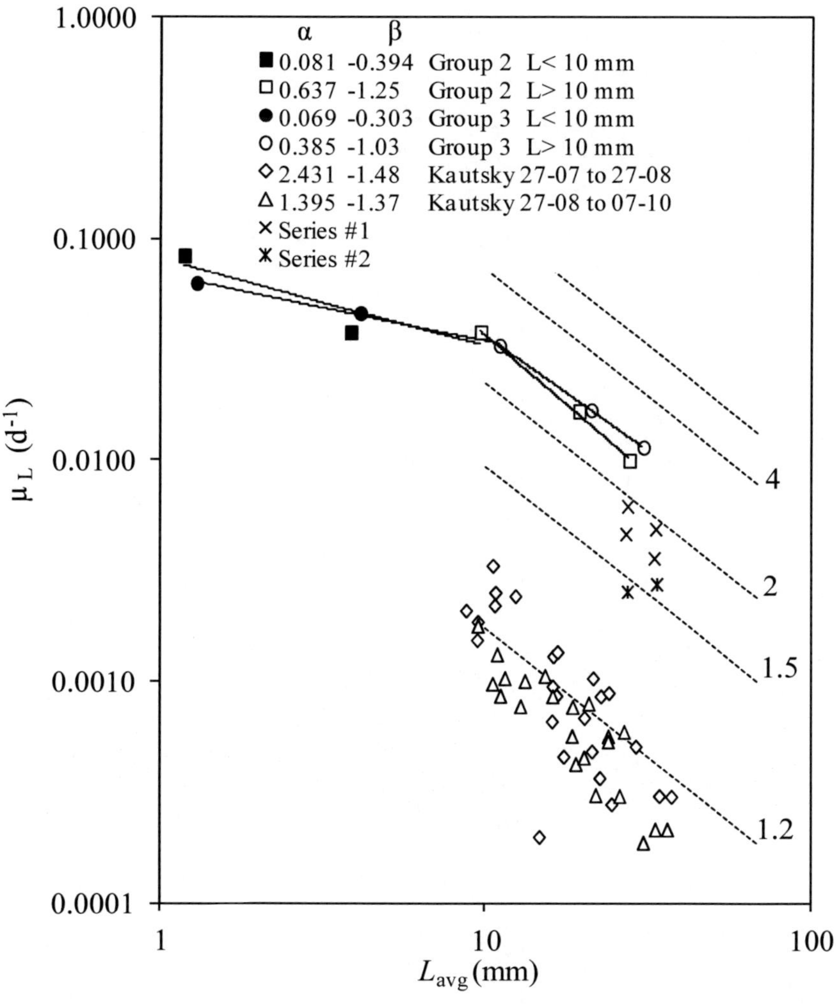

Figure 9. *Mytilus edulis* (farm-ropes: Groups 2 and 3; net-bags: Series #1 and #2). Shell length specific growth rate (μ_L, Eq.7) versus average shell length (L_{avg}, Eq.10) over each successive growth interval, cf. Table 1. Coefficients of power-law relations for present data of shell length smaller and larger than about 10 mm, as well as for re-plotted data for $L_{avg} > 10$ mm from Kautsky (1982, Figure 5 therein) are inserted in figure. Dotted lines: growth model (Eq.12) for 5 values of chl a concentration ($\mu g\ L^{-1}$).

Table 3. ***Mytilus edulis*** **(net-bags, Series #1a, #1b and Series #2)**

Series # Date 2011	Δt (d)	Chl *a* (μg L^{-1})	T (°C)	S (psu)	L_0 (mm)	L_{end} (mm)	L_{avg} (mm)	μ_L (d^{-1})	W_0 (mg)	W_{end} (mg)	W_{avg} (mg)	μ (% d^{-1})	CI_0 (mg cm^{-3})	CI_{end} (mg cm^{-3})
#1a	16	1.9±0.2	13.7±1.7	19.8	25.8±0.7	28.5±0.9	27.1	0.00622	55.3±12.7	171.9±47.3	97.5	7.1	3.2±0.7	7.3±1.6
4–20 Oct					31.8±0.4	33.7±1.1	32.7	0.00363	111.8±20.8	269.2±45.1	173.5	5.5	3.5±0.7	7.0±0.8
#1b	16	1.9±0.2	13.7±1.7	19.8	25.8±0.7	27.8±1.3	26.8	0.00467	55.3±12.7	167.6±53.8	96.3	6.9	3.2±0.7	7.6±2.0
4–20 Oct					31.8±0.4	34.4±0.8	33.1	0.00491	111.8±20.8	301.0±46.0	183.5	6.3	3.5±0.7	7.4±1.0
#2	13	3.3±1.1	9.4±1.6	12.8	26.6±1.2	27.5±0.8	27.0	0.00256	92.2±17.0	163.3±22.3	122.7	4.2	4.9±0.6	7.9±0.8
10–23 Nov					32.7±0.5	33.9±0.5	33.3	0.00277	199.6±41.4	309.1±49.7	248.4	3.2	5.7±1.2	8.0±1.2

Duration (Δt) of mussel growth in net-bags at MarBioShell mussel farm, mean (±S.D.) values of chlorophyll *a* (chl *a*), temperature (T), salinity (S), initial shell length (L_0), body dry weight (W_0), initial condition index (CI_0, Eq.2), final shell length (L_{end}), final body dry weight ($W_{end,}$), final condition index (CI_{end}), average shell length and body dry weight (L_{avg} and W_{avg} from Eq.10 and 4), and average shell length and weight specific growth rates (μ_L from table values and μ from slopes in Figures 10a and 10b). n = 10-17. Chl *a*, temperature and salinity data are obtained from the Danish Nature Agency.

DISCUSSION

Mussel-Ropes

For Group 3 of 10 biggest mussels on farm-ropes, results of the specific growth rate during the growth season in Figure 5 are seen to follow the power-law ($\mu = aW^b$) quite well in the two size ranges: for small farm-rope mussels ($W_{avg} < 10$ mg) with exponent $b = -0.13$, which is close to the suggested value of $b = -0.1$, and for larger farm-rope mussels ($W_{avg} > 10$ mg) with exponent $b = -0.25$, which is somewhat larger than $b = -0.34$ predicted by the BEG model. However, it is noted that the dotted lines of the growth model in Figure 5 indicate a predicted ambient chl *a* concentration of about 3 to 4 µg L^{-1} in fair agreement with the general level of observed average chl *a* concentration (3.1±2.1 µg L^{-1}, Figure 2) at the site during the period. Statistical treatment (PPMC) reveals that slopes describing the correlation between dry weight of soft parts and weight specific growth rate are significantly different from zero (Group 1: $r_p = -0.999$, N = 3, P = 0.0284; Group 2: $r_p = -0.999$, N = 3, P = 0.0216), and the slopes are significantly different between mussel groups (ANCOVA, $F_{(1,3)} = 0.243$, P = 0.656).

It is interesting to note that the correlation of data from a growth experiment with *Mytilus edulis* of $L > 10$ mm kept one year in cages could be expressed as: $W(\mu g) = 8.90L(mm)^{2.92}$ by Kautsky (1982, Figure 17 therein) which is near identical to that of Eq.(13) as shown in Figure 8. The same is true for data of Stirling and Okumuş (1994, re-plotted from their Figures 3 and 4) on year old rope-grown blue mussels > 10 mm on the west coast of Scotland at Loch Etive (LE) and Loch Leven (LL) during May-Sept (having regression lines of $W = 61.24L^{2.45}$ and $W = 3.04L^{3.26}$, respectively). Later in the season, however, data fall below these regression lines as the chl *a* concentration decreases to values well below 0.6 µg L^{-1}. (To convert the "ash-free dry meat weight" given by Stirling and Okumuş (1994) into dry weight of soft parts we have multiplied by a factor 1.1; own determination of conversion factor). Our previous results (Riisgård et al. 2011b, Figure 12 therein) on mussels ($16<L<83$ mm) collected in Kerteminde Fjord, Denmark, were correlated by $W = 6.0L^{3.0}$ which is close to the foregoing correlations, confirming the general trend of dry weight of soft parts being approximately proportional to L^3 for $L > 10$ mm. Further, equating Eqs.(1) and (13) shows that the 2 regression lines in Figure 8 intersect at $L = 12.1$ mm, corresponding to $W = 10.3$ mg in good agreement with the change-over observed in Figure 5. These results confirm that increase of body dry weight and shell length during the ontogeny of *M. edulis* changes character around $W \approx 10$ mg and $L \approx 10$ mm. However, these allometric transitions are probably smooth and they remain to be explained biologically.

The weight specific growth rate µ is a useful quantity because it does not explicitly involve absolute time but only time intervals to give the instantaneous rate of growth for a given size W of mussel. Knowing $\mu(W)$, however, the time history of growth from a given initial size (W_1 at time t_1) can be obtained by integration, which becomes simple (Eq.6) when $\mu(W)$ follows a power law as suggested by the BEG model. The same is true for growth in terms of shell length where the data supported correlation of form $W = cL^d$ (Figure 8) for small and large mussels transforms the power law of the BEG model ($\mu = aW^b$) into a power law for shell length specific growth rate (μ_L, Eq.12). The present μ_L-values for Groups 2 and 3 (Table 1) are plotted versus L_{avg} in Figure 9. Inspection indicates that both groups follow power-law trends with exponents $\beta = -0.394$ and -0.303, respectively, for shell length < 10 mm, and exponents $\beta = -1.25$ and -1.03, respectively, for shell length > 10 mm, the correlation coefficients of the two data sets being 0.99 and 1.00, respectively. This means that the growth of the 10 biggest mussels is quite representative of the biggest 20% of all of the population, and we used thc averages of exponents $\beta = -0.352$ and 1.12, respectively, and the averages of coefficient $\alpha = 0.0755$ and 0.480, respectively (obtained from regression to the 2 groups of data jointly) to predict growth by Eq.(9) for $L < 10$ and > 10 mm, respectively, in preparing Figure 9. Statistical treatment (PPMC) reveals that slopes describing the correlation between shell length and shell length specific growth rate are significantly different from zero for both groups of mussels < 10 mm ($r_p = -0.861$, N = 6, P = 0.028) and > 10 mm ($r_p = -0.901$, N = 6, P = 0.014) shell length. Slopes of regression lines between groups are significantly different from each other (ANCOVA, $F_{(1,9)} = 0.623$, P = 0.450).

In general, μ_L of Eq.(8) may prove to be a useful relation for studies of mussel growth if only measured in terms of shell length. It was used by Kautsky (1982, his Figure 5 and Table 1, in a form based on $\log_{10}L$ in place of $\ln L$) to study cage-growth of mussels in the brackish Baltic Sea near the Askö Laboratory during different seasons. Some of these results for $L >$ 10 mm (appropriately converted into the present μ_L-values) have been re-plotted in Figure 9 and appear to be correlated by Eq.(8) with β-values close to those of the present data, but the magnitude of growth (α-value) is 10 to 20 times lower than found for mussels on farm-ropes in the Great Belt in the present study. The slow growth may however be explained by a low salinity of around 7 psu at the growth site on the Swedish Baltic coast and a low mean annual chlorophyll *a* concentration of approximately 1.5 μg L^{-1} (Kautsky 2008). More specifically according to Ulf Larsson (personal communication), the mean annual chl *a* for the Askö waters was 1.56 μg L^{-1} for the 34 year period from 1976 to 2009 with variations from 1.11 μg L^{-1} (1989) to 2.46 μg L^{-1} (2008, sole year with chl *a* > 2 μg L^{-1}) measured on water samples collected at least 20 times per year in the upper 20 m (cf. the Swedish data base SMHI, www.smhi.se). For this reason the model predictions from Eq.(12) have also been shown in Figure 9 (dotted lines) for a number of values of chl *a* which suggest that the chl *a* concentration might have been as low as 1.2 μg L^{-1} provided feeding conditions were otherwise close to optimal and no effect of low salinity. Other data (Kautsky 1982, Figure 22 therein) confirm the slow growth, taking of the order of 10 to 12 years to reach a shell length of 30 mm as compared to about one year at the present Great Belt site. Growth rate and maximum size of *Mytilus edulis* is much lower in the brackish Baltic Sea (7 psu) than in the high saline North Sea (28 psu), but reciprocally transplanted mussels grew at rates similar to those of native mussels at each site (Kautsky et al. 1990), indicating that the observed variations in growth may be explained by differences in salinity and chl *a*. Other observations of shell growth rates from the northern Baltic Sea (4.5 psu) include Westerbom et al. (2002, Figure 6 therein): 12 years to reach 30 mm with max. 2.9 mm y^{-1}; Vuorinen et al. (2002, Figure 6 therein): 8 years to reach 30 to 35 mm. For comparison, Schütz (1964) states that mussels at the Kiel Bight (18 psu) attain a shell length of 30 mm during the first year, as also found in the present study (16 psu, Figure 2). We have also calculated $\mu_L(L_{avg})$ for data from Stirling and Okumus (1994, their Series LE and LL) and when plotted in Figure 9 (not shown) the results scatter around the present data shown from Series #1 and #2 near $C = 1.5$ to 2 μg L^{-1} according to the BEG-model in accord with the low chl *a* concentrations given by Stirling and Okumus (1994). However, it should be noted that spawning cycle and other factors may lead to changes in the relationship between shell length (L) and dry weight of soft parts (W) (e.g. Dare 1976), hence of the condition index ($CI = W/L^3$), as seen from the variation of such values in Table 3.

Net-Bags

Series #1a and #1b. The weight specific growth rates were high both within the mussel farm, between 5.5 and 7.1 % d^{-1} for 31.8 and 25.8 mm shell length mussels, respectively (Table 3, Figure 10), and at the outer edge of the farm, 6.3 and 6.9 % d^{-1} for 31.8 and 25.8 mm mussels, respectively, compared to a previous study in the same area (Riisgård et al. 2012a). This suggests optimal conditions with no negative effects of e.g. high current speeds or changing salinities during the growth period and that the mean chl *a* concentration in the surrounding water must have been relatively high during the growth period, which according to the growth model shown in Figure 5 should have been around 5 μg L^{-1}, but the actually measured mean concentration from October 2011 was low, about 1.9±0.2 μg L^{-1} (Figure 2, Table 3). Further, Figure 5 (data C and D) shows that there is no significant difference between growth of net-bag mussels in the middle (#1a) or in the outer region (#1b) of the farm site, and the higher growth rates than those of farm-rope mussels of the same size could be a result of less intraspecific competition for nutrition (Fréchette et al. 1992).

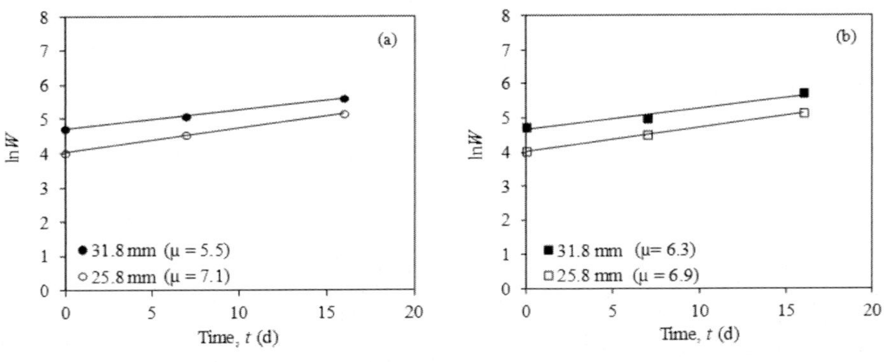

Figure 10. *Mytilus edulis* (net-bags, Series #1a, #1b). Growth data in terms of ln$W$ during the period 4 to 20 October 2011. Series #1a (a) are mussels suspended within the mussel farm, and Series #1b (b) are mussels suspended at the outer edge of the mussel-farm. μ (% d^{-1}) shows the average weight specific growth rate of the 2 size classes in the growth period.

Series #2. The weight specific growth rates were 4.2 and 3.2 % d^{-1} for 26.6 and 32.7 mm shell length mussels, respectively (Table 3, Figure 11), which according to the dotted lines of the growth model in Figure 5 indicate a predicted ambient chl *a* concentration of about 3.5 μg L^{-1} in good agreement with the actual concentration from November 2011 of 3.3±1.1 μg L^{-1} (Figure 2, Table 3). μ_L-values in Figure 9, on the other hand, suggest the much lower chl *a* value of $C \sim 1.5$ μg L^{-1}.

These inconsistencies may be a result of seasonal differences in growth between Series #1 (4 to 20 Oct) and Series #2 (10 to 23 Nov) such that Series #2 falls outside of the favorable growth period intended for study, just as the last data points of Tables 1 and 2 were ignored.

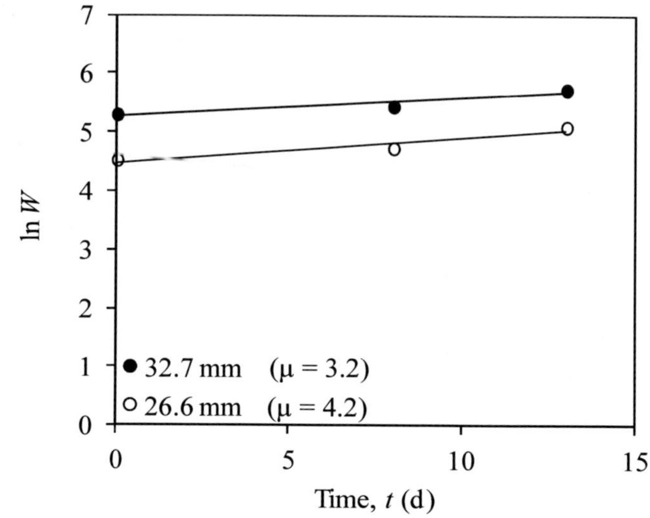

Figure 11. *Mytilus edulis* (net-bags, Series #2). Growth data in terms of ln*W* of mussels suspended at the outer edge of the mussel-farm during the period 10 to 23 November 2011. μ (% d^{-1}) shows the average weight specific growth rate of the 2 size classes in the growth period.

It is notable that the calculated values of shell length specific growth rates μ_L for Series #1 and #2 shown in Figure 9 are low compared to the shell growth of Groups 2 and 3 and that this is not the case for the weight specific growth rates μ of these groups in Figure 5. This phenomenon may probably be explained by delayed growth of shells compared to somatic growth of net-bag mussels with an initially low condition index, cf. about doubling of *CI* in the field-growth period (Table 3).

CONCLUSION

The present growth data of juvenile mussels on ropes, supplemented by growth data of mussels in suspended net-bags at the same site, support the bioenergetic growth model of Eq.(5) for $W > 10$ mg ($b = -0.34$) and the modified form for $W < 10$ mg ($b \approx -0.13$), covering more than 4 decades of size in terms of dry weight of soft parts (*W*) as shown in Figure 5. From the growth data on shell length (*L*), leading to Eq.(13) to supplement Eq.(1) and leading to Figure 9, it has been experimentally confirmed that increase of body dry weight (somatic growth) and shell length during the ontogeny of the blue mussel changes character around $W \approx 10$ mg and $L \approx 10$ mm. Further, using Eqs.(1) and (13), the somatic growth model has been expressed in terms of shell length as Eq.(12) and used to show that other field data on mussel growth in the Baltic Sea (Kautsky 1982) lends well to interpretation by the model. The explicit model dependence on chlorophyll *a* concentration is well established for larger mussels ($W > 10$ mg) for which model predictions have been compared to actual growth rates in the field, but similar predictions remains to be established for smaller mussels ($W < 10$ mg), along with other environmental parameters such as salinity and current velocity. It is well known that e.g. Dutch and English fishermen transplant small mussels stagnant in growth from the intertidal level to below the tidal level where the mussels quickly grow to marketable sizes (Baird 1966). Because the potential for growth "remains intact for many years in mussels that have been prevented from exploiting it owing to unfavorable environmental conditions" (Jørgensen 1976) the present statement that allometric transitions take place around $W \approx 10$ mg and $L \approx 10$ mm during the ontogeny of *Mytilus edulis* is most likely universal, and not restricted to first year growth of juvenile (young) mussels during the productive season.

ACKNOWLEDGMENTS

This work formed part of the MarBioShell project supported by the Danish Agency for Science, Technology and Innovation for the period January 2008 to December 2012. Thanks are due to Dr. Daniel Pleissner for helpful assistance with statistics, to Prof. Ragnar Elmgren for valuable information, and to Lene Simonsen and Maria Ángeles Provencio López for technical assistance.

REFERENCES

Baird RH (1966) Factors affecting the growth and condition of mussels (*Mytilus edulis* L.). *Fish Invest Ser II* 25, No 2, 1−33.

Bayne BL (1965) Growth and the delay of metamorphosis of the larvae of *Mytilus edulis*. *Ophelia* 2:1-47.

Bayne BL, Widdows J (1978) The physiological ecology of two populations of *Mytilus edulis* L. *Oecologia* (Berl) 37:137−162.

Bayne BL, Worrall CM (1980) Growth and production of mussels *Mytilus edulis* from two populations. *Mar Ecol Prog Ser* 3:317−328.

Beadman HA, Willows RI, Kaiser MJ (2002) Potential applications of mussel modeling. *Helgoland Mar Res* 56:76−85.

Bourlès Y, Alunno-Bruscia M, Pouvreau S, Tollu G, Leguay D, Arnaud C (2009) Modelling growth and reproduction of the Pacific oyster *Crassostrea gigas*: Advances in the oyster-DEB model through application to a coastal pond. *J Sea Res* 62:62−71.

Brown JH, Gillooly JF, Allen AP, Savage VM, West GB (2004) Toward a metabolic theory of ecology. *Ecol* 85:1771−1789.

Clausen I, Riisgård HU (1996) Growth, filtration and respiration in the mussel *Mytilus edulis*: no regulation of the filter-pump to nutritional needs. *Mar Ecol Prog Ser* 141:37−45.

Dare PJ (1976) Settlement, growth and production of the mussel, *Mytilus edulis* L., in Morecambe Bay, England. Fish Invest Ser II 28, No 1, 1−25.

Dolmer P (1998) Seasonal and spatial variability in growth of *Mytilus edulis* L. in a brackish sound: comparisons of individual mussel growth and growth of size classes. *Fish Res* 34: 17−26.

Dolmer P, Stenalt E (2010) The impact of the adult blue mussel (*Mytilus edulis*) population on settling of conspecific larvae. *Aquacult Int* 18:3−17.

Duarte P, Fernández-Reiriz MJ, Filgueira R, Labarta U (2010) Modelling mussel growth in ecosystems with low suspended matter loads. *J Sea Res* 64:273−286.

Filgueira R, Rosland R, Grant J (2011) A comparison of scope for growth (SFG) and dynamic energy budget (DEB) models applied to the blue mussel (*Mytilus edulis*). *J Sea Res* 66:403-410.

Fréchette M, Aitken AE, Pagé L (1992) Interdependence of food and space limitation of a benthic suspension feeder: consequences for self-thinning relationships. *Mar Ecol Prog Ser* 83:55–62.

Gosling E (2003) Bivalve molluscs. Biology, ecology and culture. Fishing News Books, Blackwell, p 443.

Hamburger K, Møhlenberg F, Randløv A, Riisgård HU (1983) Size, oxygen consumption and growth in the mussel *Mytilus edulis*. *Mar Biol* 75:303–306.

Hawkins AJS, Fang JG, Pascoe PL, Zhang JH, Zhang XI, Zhu MY (2001) Modelling short-term responsive adjustments in particle clearance rate among bivalve suspension-feeders: separate unimodal models of seston volume and composition in the scallop *Chlamys farreri*. *J Exp Mar Biol Ecol* 261:61–73.

Jespersen H, Olsen K (1982) Bioenergetics in veliger larvae of *Mytilus edulis* L. *Ophelia* 21:101–113.

Jørgensen CB (1976) Growth efficiencies and factors controlling size in some mytilid bivalves, especially *Mytilus edulis* L.: review and interpretation. *Ophelia* 15:175–192.

Jørgensen CB (1981) Mortality, growth, and grazing impact of a cohort of bivalve larvae, *Mytilus edulis* L. *Ophelia* 20:185–192.

Jørgensen CB (1990) Bivalve filter feeding: hydrodynamics, bioenergetics, physiology and ecology. Olsen & Olsen, Fredensborg, Denmark. 140 pages.

Kautsky N (1982) Growth and size structure in a Baltic *Mytilus edulis* population. *Mar Biol* 68:117–133.

Kautsky H (2008) Askö area and Himmerfjärden. In: Schiewer U (ed) Ecology of Baltic coastal waters. Ecological studies 197. Springer-Verlag Berlin Heidelberg, pp 335–360.

Kautsky N, Johannesson K, Tedengren M (1990) Genotypic and phenotypic differences between Baltic and North Sea populations of *Mytilus edulis* evaluated through reciprocal transplantations. I. Growth and morphology. *Mar Ecol Prog Ser* 59:203-210.

Kristensen PS, Lassen H (1997) The production of relaid blue mussels (*Mytilus edulis* L.) in a Danish fjord. ICES *J Mar Sci* 54:854-865.

Møhlenberg F, Riisgård HU (1979) Filtration rate, using a new indirect technique, in thirteen species of suspension-feeding bivalves. *Mar Biol* 54:143–148.

Okumus I, Stirling HP (1994) Physiological energetics of cultivated mussel (*Mytilus edulis*) populations in two Scottish west coast sea lochs. *Mar Biol* 119:125–131.

Riisgård HU (1998) No foundation of a "3/4 power scaling law" for respiration in biology. *Ecol Lett* 1:71–73.

Riisgård HU, Egede PP, Saavedra IB (2011a) Feeding behaviour of mussels, *Mytilus edulis*: new observations, with a mini-review of current knowledge. *J Mar Biol* (published online doi: 10.1155/2011/312459).

Riisgård HU, Jørgensen BH, Lundgreen K, Storti F, Walther JH, Meyer KE, Larsen PS (2011b) The exhalant jet of mussels *Mytilus edulis*. *Mar Ecol Prog Ser* 437:147–164.

Riisgård HU, Lundgreen K, Larsen PS (2012a) Field data and growth model for mussels *Mytilus edulis* in Danish waters. *Mar Biol Res* 8: 683-700.

Riisgård HU, Pleissner D, Lundgreen K, Larsen PS (2012b) Growth of mussels *Mytilus edulis* at algal (*Rhodomonas salina*) concentrations below and above saturation levels for reduced filtration rate. *Mar Biol Res* (in press).

Riisgård HU, Randløv A, Kristensen PS (1980) Rates of water processing, oxygen consumption and efficiency of particle retention in veligers and young post-metamorphic *Mytilus edulis*. *Ophelia* 19:37−47.

Rosland R, Strand Ø, Alunno-Bruscia M, Bacher C, Strohmeir T (2009) Applying dynamic energy budget (DEB) theory to simulate growth and bio-energetics of blue mussels under low seston conditions. *J Sea Res* 62:49−61.

Saraiva S, van der Meer J, Kooijman S, Sousa T (2011) Modelling feeding processes in bivalves: A mechanistic approach. *Ecol Modell* 222:514−523.

Savage RE (1956) The great spatfall of mussels in the River Conway Estuary in Spring 1940. Fish Invest Ser II London 20:1−21.

Schütz L(1964) Die tierische Besiedlung der Hartböden in der Schweninemündung. Kieler Meeresforsch 20:198−217.

Seed R (1976) Ecology. In: Bayne BL (ed) Marine mussels: Their ecology and physiology, Cambridge University Press, Cambridge, pp 13-65.

Sprung M (1984) Physiological energetics of mussel larvae (*Mytilus edulis*). I. Shell growth and biomass. *Mar Ecol Prog Ser* 17:283−293.

Stirling HP, Okumus I (1994) Growth, mortaility and shell morphology of cultivated mussel (*Mytilus edulis*) stocks cross-planted between two Scottish sea lochs. *Mar Biol* 119:115−123.

Theisen BF (1975) Growth parameters of *Mytilus edulis* L. (Bivalvia) estimated from ragging data. Meddr fra Danm Fisk- og Havunders 7:99−109.

Thorson G (1961) Length of pelagic larval life in marine bottom invertebrates as related to larval transport by ocean currents. In: Sears M (ed) Oceanography. Washington DC, AAAS, pp 455−474.

van der Meer J (2006) Metabolic theories in ecology. *Trends Ecol Evol* 21:136−140.

van der Veer HW, Cardoso JFMF, van der Meer J (2006) The estimation of DEB parameters for various Northeast Atlantic bivalve species. *J Sea Res* 56:107−124.

Vuorinen L, Antsulevich E, Maximovich NV (2002) Spatial distribution and growth of the common mussel *Mytilus edulis* L. in the archipelago of SW-Finland, northern Baltic Sea. *Boreal Envir Res* 7:41−52.

Westerbom M, Kilpi M, Mustonen O (2002) Blue mussels, *Mytilus edulis*, at the edge of the range: population structure, growth and biomass along a salinity gradient in the north-eastern Baltic Sea. *Mar Biol* 140:991−999.

Widdows J, Donkin P, Brinsley MD, Evans SV, Salkeld PN, Franklin A, Law RJ, Waldock MJ (1995) Scope for growth and contaminant levels in North Sea mussels *Mytilus edulis*. *Mar Ecol Prog Ser* 127:131−148.

In: Mussels: Ecology, Life Habits and Control
Editors: Jarek Nowak and Michal Kozlowski
ISBN: 978-1-62618-083-3

Chapter 5

THE SYNTHESIS OF VITELLOGENIN IN THE MUSSEL MYTILUS GALLOPROVINCIALIS: AN INTERDISCIPLINARY APPROACH

Marisa Agnese, Marina Prisco, Luigi Rosati, Mariailaria Verderame, Enrico De Meo, Ermelinda Limatola, Rosanna Del Gaudio, Serena Aceto and Piero Andreuccetti*

Department of Biology,
University of Naples Federico II, Naples, Italy

ABSTRACT

Vitellogenins (VTGs) represent the most important precursors of vitellins (Vns), the egg yolk proteins utilized by oviparous and ovoviviparous animals to guarantee the embryo development. In mature females, VTGs are synthesized in different districts under the control of sexual hormones, as 17β-estradiol, and transferred to the growing oocytes via the bloodstream. In the oocytes, VTGs are processed by cathepsins into Vns that are then stored into the yolk platelets. Differently, in immature females and in the males VTGs are not synthesized, but their synthesis may be activated by xeno-estrogens, an heterogeneous group of environmental contaminants that modify the endocrine reproductive system with consequent production of molecules not present physiologically, as VTG. For this reason, VTG production, in immature females and males, is an useful biomarker to study the health of aquatic and terrestrial ecosystems. Till now, the VTG was utilized as biomarker mostly in fish, and more recently in mussels too. Such organisms indeed have acquired a prominent role as sentinel species in aquatic pollution-monitoring programs. Differently from fish, the knowledge of VTG synthesis and its endocrine control in the mussels is scant. At this regard we carried out an interdisciplinary investigation to identify the site/s of VTG synthesis in the mussel *Mytilus galloprovincialis*, that may represent a useful sentinel species in the Mediterranean sea. In this organism we have investigated two districts, the ovary and the

* Correspondence should be addressed to: Piero Andreuccetti, c/o Department of Biology, Via Mezzocannone 8, 80134 NAPOLI, Italy; Phone 0039 081 2535176; Fax 0039 081 2535035; E-mail to: piero.andreuccetti@unina.it.

digestive gland, that are the most probably sites of VTG synthesis in the mussels. Results of RT-PCR, in situ hybridization, electron-microscopy and immunohistochemistry experiments, taken together, provide evidence that VTG synthesis occurs only in the ovary of females, and particularly both in the oocyte (autosynthesis) and follicle and adipogranular cells (eterosynthesis). The possibility that in the mussel *Mytilus galloprovincialis* VTG may be used as biomarker in ecotoxicology is also discussed.

INTRODUCTION

Marine pollution is a problem of great relevance, especially in Mediterranean sea; the biomonitoring of environmental pollution is carried out using appropriate biomarkers. Among the biomarkers of xenoestrogens pollution, the vitellogenin (VTG) is the elected one. VTG is a lipophosphoglicoprotein highly represented in the eggs of oviparous and ovoviviparous vertebrates and it is the precursor of most of the yolk proteins that accumulate as storage material in the oocyte during the oocyte growth. VTG is considered a stage- and sex-specific protein; indeed, in vertebrates it is synthesized in the liver and/or in the ovary of mature females under the control of sexual hormones, as 17β-estradiol, while it is not present in immature females and in males (Wallace, 1985; Polzonetti-Magni et al., 2004; Prisco et al., 2004). Once synthesized, VTG is transferred via the bloodstream to the growing oocytes, where it is processed by cathepsins into vitellins (Vns) that are then stored into the yolk platelets (Wallace, 1985; Polzonetti-Magni et al., 2004). VTG is not physiologically present in immature females and in males, but its synthesis can be induced in a dose-dependent manner by xeno-estrogens; this feature makes this protein a useful biomarker of environmental pollution, being largely used as a biomarker, especially in fish (Maradonna and Carnevali, 2007).

In the last years, the mussels, as fish, are becoming important sentinel species in biomonitoring studies, due to their peculiar features: these animals are sessile and filter-feeding organisms, show a wide distribution along most coastal zones in the world, and have a high bioaccumulation and low biotransformation potential for both organic and inorganic contaminants (Ortiz-Zarragoitia and Cajaraville, 2006; Oehlmann and SchulteOehlmann 2002; Smolders et al. 2003). The knowledge of VTG synthesis and its endocrine control in the mussels is scant and often based on indirect methods. The sites identified as responsible for VTG synthesis are different, so that it is possible to recognize an authosynthesis, when VTG is synthesized directly within the oocyte, and an heterosynthesis, when the synthesis occurs outside the oocyte. In *Mytilus edulis*, basing on morphological investigations, it has been reported an autosynthesis (De Jon-Brink et al., 1983; Pipe, 1985). In *Cassostrea virginica* both an autosynthesis and an heterosynthesis have been reported, the latter one due to the vesicular connective tissue, surrounding the ovarian follicles (Eckelbarger and Davis, 1996). In *Patinopecten yessoensis* (Osada et al. 2003, 2004) and in *Haliotis discus hannai* (Matsumoto et al. 2008), basing on immunohistochemistry and molecular investigations, the heterosynthesis, that occurs in the follicle cells, seems responsible for VTG production. Furthermore, VTG heterosynthesis is also demonstrated by the assay of alkali-labile phosphate (ALP), an indirect technique utilized to determine the presence of VTG-like proteins (Blaise et al., 1999; Pampanin et al., 2005; Ricciardi et al., 2008) in the hemolymph of several molluscs treated with estrogen (Blaise et al., 1999; Gagnè et al., 2001) or exposed to environmental chemical known to induce changes and alterations of the endocrine functions (endocrine-disruptors) (Blaise et al., 1999; Gagné et al., 2002; Verslycke et al., 2002; Quinn et al., 2004; Marin and Matozzo, 2004; Versonnen et al., 2004; Matozzo and Marin, 2005; Pampanin et al., 2005; Aarab et al., 2004, 2006; Ortiz-Zarragoitia and Cajaraville, 2006; Zorita et al., 2006; Matozzo et al., 2008). These results underline once again the importance of VTG as a biomarker of environmental pollution.

In the Mediterranean sea the most represented species of mussels is *Mytilus galloprovincialis*, that may represent a useful sentinel species for the biomonitoring studies. However, information about VTG in this species are still very scant. The aim of this work is to amplify the knowledge of VTG synthesis sites, to acquire propedeutical information to use this protein as a biomarker. Starting from indirect observations based on electron microscopy analyses, followed by RT-PCR, in situ hybridization and immunohistochemistry investigations, we provide evidences that in this species autosynthesis and heterosynthesis coexist, as VTG synthesis occurs in the oocyte and in the follicle and adipogranular cells. We also demonstrated that the digestive gland is not involved in VTG synthesis. The possibility that in the mussel *Mytilus galloprovincialis* VTG may be used as biomarker is also discussed.

Materials and Methods

Animals

100 female specimens in different phases of the reproductive cycle (stage 0, stage 1, stage 2, stage 3 according to Duinker et al 2008) of the mussel *Mytilus galloprovincialis* were collected in the Gulf of Naples; the ovaries and digestive glands were excised using sterile RNase-free dissecting equipment and, once removed, were frozen at -80°C or fixed for 24 h in Bouin's solution and then dehydrated with graded ethanol and embedded in paraffin wax. Five μm thick sections were stained with Mallory's trichrome stain (Mazzi, 1977) or placed on Superfrost Plus or Polylisine glass slides (Menzel-Glaser, Braunschweig, Germany) for in situ hybridization and immunohistochemical experiments. Furthermore, pieces of the ovaries were prepared and used for electron microscopy, as described below.

RT-PCR

Total RNA was extracted from ovaries and digestive gland using TRI-Reagent (Sigma-Aldrich, USA) according to the manufacturer's instructions and, after the evaluation on 1% agarose gel, it was retrotranscribed with SuperScript II Reverse Transcriptase (Invitrogen) using oligo(dT)$_{15\text{-}18}$ as primers.

cDNA was used as template in PCR reaction referring to primer sequence reported by Puinean et al. (2006). The PCR was carried out for 35 cycles: 30 s at 95°C, 1 min at 58°C, 30 s at 72°C. A negative control, without cDNA, was performed. The 18S ribosomal subunit of Mytilus was used as internal control of cDNA. The PCR product was analyzed on 2% agarose gel stained with ethidium bromide for visualization; the amplified product was purified using QIAquick gel extraction kit (QIAGEN), following the manufacturer's instruction; steril water was added to elute DNA from spin column. DNA fragment was then directly sequenced with BigDye Terminator v1.1 Cycle Sequencing Kit (Applied Biosystems) using the same primers employed for the PCR amplifcations and run on the ABI PRISM 310 Genetic Analyzer (Applied Biosystems). The obtained nucleotide sequence was searched for similarity using BLASTn and aligned using ClustalW2. (Agnese et al., 2012).

In Situ Hybridization

Partial sequence of *Mytilus edulis* vitellogenin (GenBank accession number AY679116.1) cloned in pCR 2.1 TOPO vector (Invitrogen), kindly gifted by Dr.J. M. Rotchell (University of Hull HU6 7RX UK), was used for subcloning experiments in pGEM®- 3Z vector and was linearized to perform *in vitro* transcription reaction. Digoxigenin (DIG)-labeled sense and antisense *Vtg* RNA probes from *Mytilus Vtg* were synthesized using RNA labeling kit (Roche), according to manufacturer's instructions. In situ hybridization was carried out on ovaries and digestive gland sections as reported elsewhere (Agnese et al., 2012).

Transmission Electron Microscopy

For electron microscopy, samples of female gonads were immersed in Karnowsky's fixative (4% paraformaldehyde, 5% glutaraldeyde in phosphate buffer 0.1M, pH 7.4) for 2 h at room temperature. Tissues were then washed overnight in freshly prepared buffer.

Post-fixation was carried out incubating samples for 1 h at 4°C in 2% osmium tetroxide in phosphate buffer. The samples were then dehydrated through an ascending series of ethanols ending in 100% ethanol, transferred through two changes of propylene oxide, 5 min. each at room temperature, infiltrated in 1:1 (v:v) propylene oxide:Epon, and finally embedded in Epon. Semi-thin and ultra-thin sections were cut on Reichert-Jung Supernova ultramicrotome. Semithin sections were stained with toluidine blue; ultrathin sections were stained with uranyl acetate and lead citrate, and examined under a CM 12 TEM. Electron microscope investigations were carried out at the "C.I.S.M.E" of the University of Naples Federico II.

Immunohistochemistry

Immunohistochemistry was carried out as reported elsewhere (Agnese et al., 2012). The primary polyclonal antibody used was an anti-vitellogenin of *Haliotis discus hannai*, kindly gifted by Dr. Masahiko Awaji (National Research Institute of Aquaculture, Minami-Ise, Mie 516-0193, Japan).

RESULTS

Transmission Electron Microscopy

Differently from oogonia and previtellogenic oocytes, that showed a few cytoplasmic organelles and no yolk platelets (data not shown), *Mytilus galloprovincialis* early vitellogenic oocytes ovoid in shape were characterized by the presence of RER cisternae and Golgi complexes associated to electron dense granules that represent nascent yolk platelets (Figures

1A, B). At this stage no endocytotic activity was evident along the oocyte surface (Figure 1B). Vitellogenic oocytes (Figure 1A) began to assume the typical pear shape with the basal region connected to the wall of the follicle and the apical region projecting toward the lumen of the follicle.

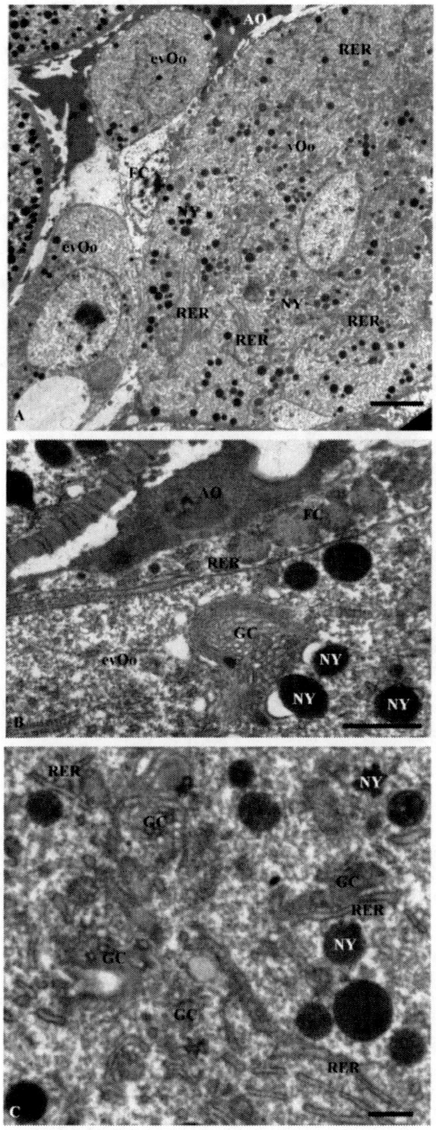

Figure 1. Early vitellogenic and vitellogenic oocytes at transmission electron microscope. A: Two early vitellogenic oocyte (evOo) and a vitellogenic oocyte (vOo) are evident. Note the well extended RER cisternae and the numerous yolk nascent platelets (NY), more represented in vitellogenic than in early vitellogenic oocytes. A follicle cell (FC) is evident. AO = Atretic Oocyte. Bar = 10 µm. B: An early vitellogenic oocyte (evOo) nest to a follicle cell (FC): no endocytotic activity is evident along the oocyte membrane. Note also the presence of RER within the follicle cell. An extended Golgi complex (GC) nest to nascent yolk platelets (NY) is detectable. AO = Atretic Oocyte. Bar = 5 µm C: Ooplasm of a vitellogenic oocyte: note the numerous RER cisternae, Golgi complexes and the presence of electron dense material within cisternae as well as the presence of nascent yolk platelets (NY) as it suggests the irregular bordering. Bar = 5 µm.

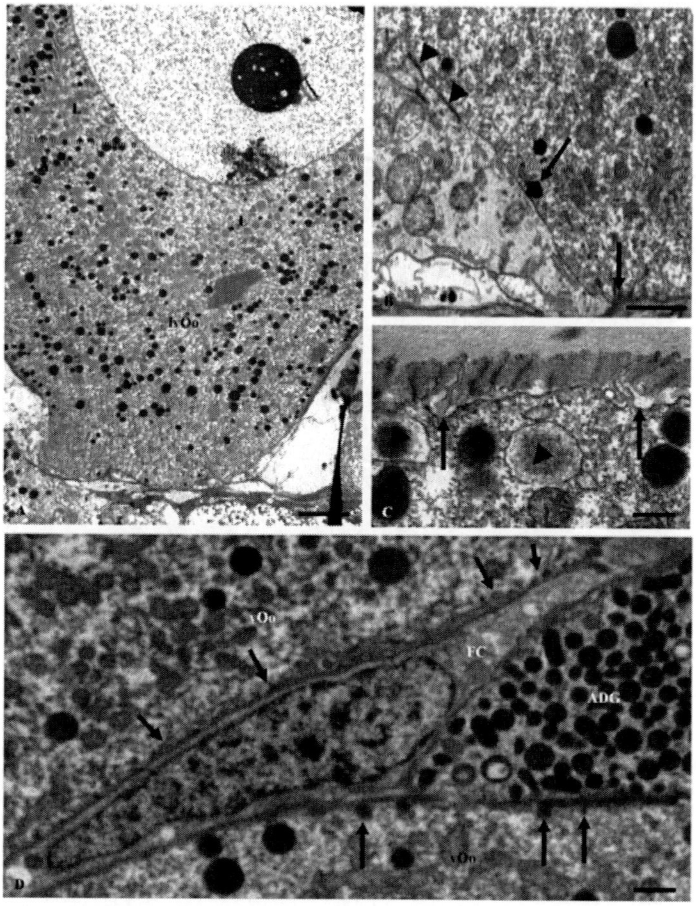

Figure 2. Late vitellogenic oocytes peduncolated in shape. A: Late vitellogenic oocyte (lvOo): the ooplasm is characterized by a large amount of yolk platelets (Y) and lipid droplets (L). Bar = 10 μm B: the oocyte cortex shows endocytotic pictures (arrows); electron dense material (arrowheads) is present in the intercellular space between the oocyte and follicle cell. Bar = 10 μm C: endocytotic activity by oocyte: surface invaginations (arrow) and small vacuoles (arrowheads) containing faintly electron dense materials are evident in the oocyte cortex. Bar = 2 μm D: A follicle and an adipogranular cell: an endocytotic activity is evident along their surface (arrows).
Bar= 2 μm.

These oocytes were characterized by an extended RER and Golgi complexes connected to nascent yolk platelets irregular in shape for the presence of merging vesicles (Figures 1B, C). A few follicle cells (Figure 1A) containing an extended RER and Golgi complex (Figure 1B) were evident.

Finally, late vitellogenic oocytes were peduncolated in shape, being anchored to the ovarian follicle wall with the basal region (Figure 2A). The ooplasm was characterized by the presence of RER cisternae and Golgi complexes, and by quite numerous yolk platelets and lipid droplets (Figure 2A). In vitellogenic and late vitellogenic oocytes, differently from the early vitellogenic oocytes, a quite high endocytotic activity was evident along the oocyte surface (Figures 2B, C), particularly in the area of contact between the oocyte and the follicle and adipogranular cells (Figure 2D).

RT-PCR

The expression levels of mussel VTG gene were investigated by RT-PCR experiments. Results of these experiments, showed in Figure 3, were cDNA fragments of about 130 bp in length, indicating that Vtg gene is transcribed in mussels growing oocyte.

It is evident that, during the gonad developmental stages, at different stages of oocyte growth, *Vtg* transcripts are differentially detected: *Vtg* is expressed only in the gonads containing oocytes in stages 1-2, i.e. in vitellogenesis (Figure 3B); unlike, *Vtg* is not expressed in the gonads in stage 0, i.e. in previtellogenesis (Figure 3A), and in stage 3, i.e. with full grown oocytes (Figure 3C). Furthermore, *Vtg* is never expressed in the digestive gland, independently from the oocyte stage (Figure 3A-C). As expected, negative PCR control showed no amplification band (Figure 3D). In addition, results of BLASTn (NCBI Data Bank) analyses of the obtained nucleotide sequence confirmed a high matching (97%) with the known *Vtg* sequence of *Mytilus edulis*.

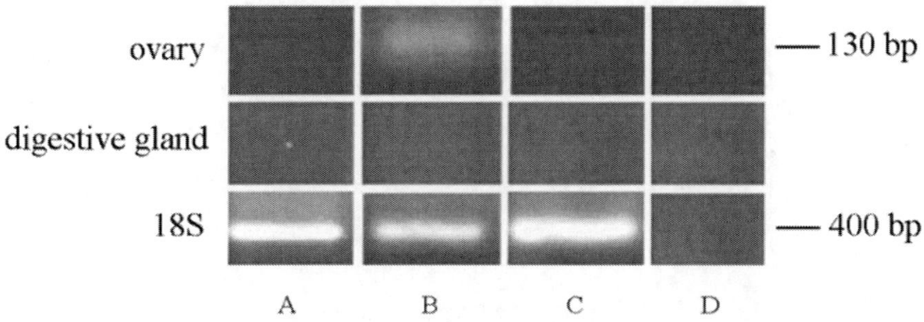

Figure 3. RT-PCR evidence of *Vtg* expression in the ovary during the different phases of oocyte growth.and digestive gland. The amplification product of about 130 bp is evident only in the gonad at the stages 1-2 (vitellogenesis) (B); no signal is evident in stage 0 (previtellogenesis) (A) and 3 (full grown oocytes) (C) as well as in the digestive gland, independently from the stage of the oocyte growth (A-C). Negative control was obtained by omitting cDNA (D). The 18S ribosomal RNA is the quality control of cDNA.

Localization of *VTG* mRNA in Ovarian Follicles and in Digestive Gland

In situ hybridization results revealed that *Vtg* mRNA was widely transcribed in the ovary (Figure 4A-E), while it was not expressed in the digestive gland (Figure 4G). In the ovary, the signal was evident in both germ and somatic cells (Figure 4B, D). A positive signal was evident in early and in vitellogenic oocytes (Figure 4B, D), while it did not occur in full grown oocytes (Figure 4E). The hybridization signal was present also in follicle cells (Figure 4B, D) and in the adipogranular cells that are present in the connective tissue among ovarian follicles (Figure 4B). No background signal was observed in the sense control experiments performed on adjacent sections to those showed in panel A-E hybridized with antisense *Vtg* probe (Figure 4E, inset). As concerns the digestive gland, that is constituted by numerous blind-ending tubules (Figure 4F), no hybridization signal was present independently from the organization of the ovarian follicles present within the gonad of the same female specimen (Figure 4G).

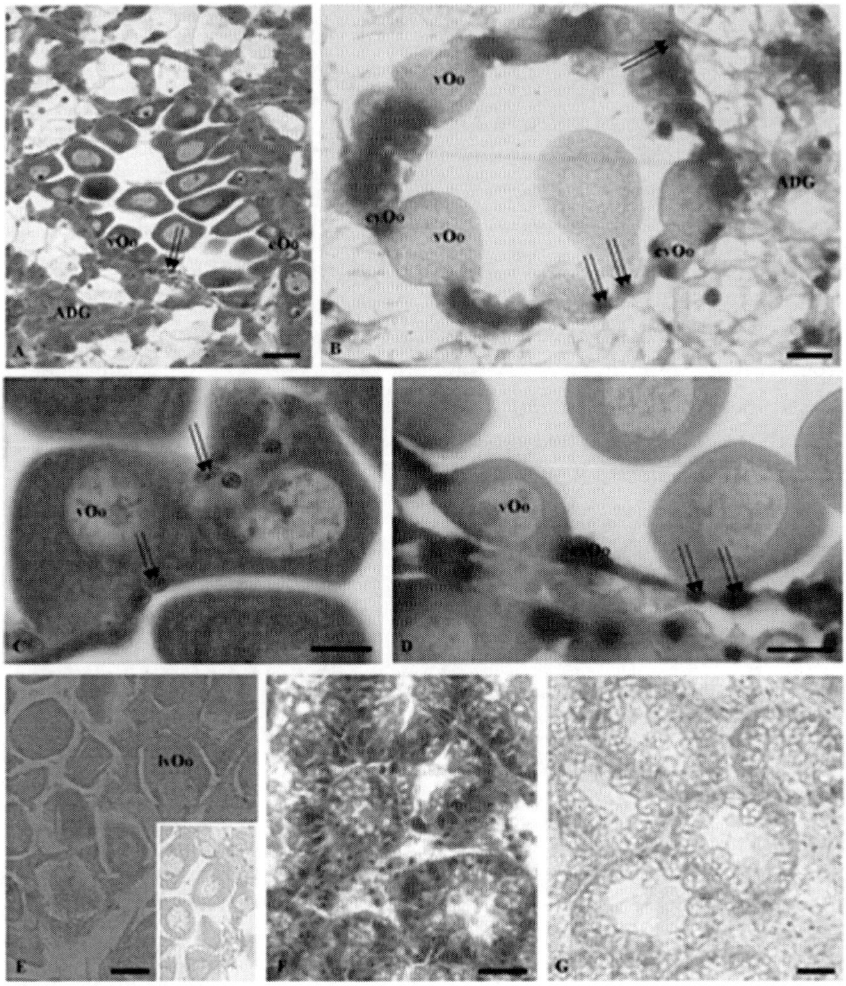

Figure 4. Localization of VTG mRNA in the ovary (A-E) and in digestive gland (G) of *Mytilus galloprovincialis* by in situ hybridization. The hybridization signal, that appears as blue areas, is evident in the germ and somatic cells of the ovary (B, D); particularly, it is present in early (evOo) and in vitellogenic oocytes (vOo) (B, D), while it is not present in full grown oocytes (lvOo) (E). The hybridization signal is present also in follicle cells (double arrows, B, D) and in adipogranular cells (ADG) that are present in the connective tissue among ovarian follicles (B, D). No signal was evident in the control sections (E, inset) and in the digestive gland (G). The Mallory trichrome stain is reported for ovary and digestive gland to put in evidence their organization (A, C, F). Bars = 30 µm.

Immunohistochemistry

Immunohistochemistry results revealed that VTG was widely distributed in the ovary (Figs. 5A, B), while it was not present in the digestive gland (Figure 5C). In the ovary VTG was distributed in both germ and somatic cells (Figs. 5A, B). A positive signal was present in vitellogenic (Figure 5A) and full grown (Figure 5B) oocytes, where the signal is located particularly in the juxta-nuclear area (Figure 5B). The immunohistochemical signal was observed also in follicle and adipogranular cells (Figures 5A, B), while no signal was evident within vesicular cells (Figure 5A). No signal was evident within digestive gland (Figure 5C), as well as in the control sections performed omitting the primary antibody (Figure 5D).

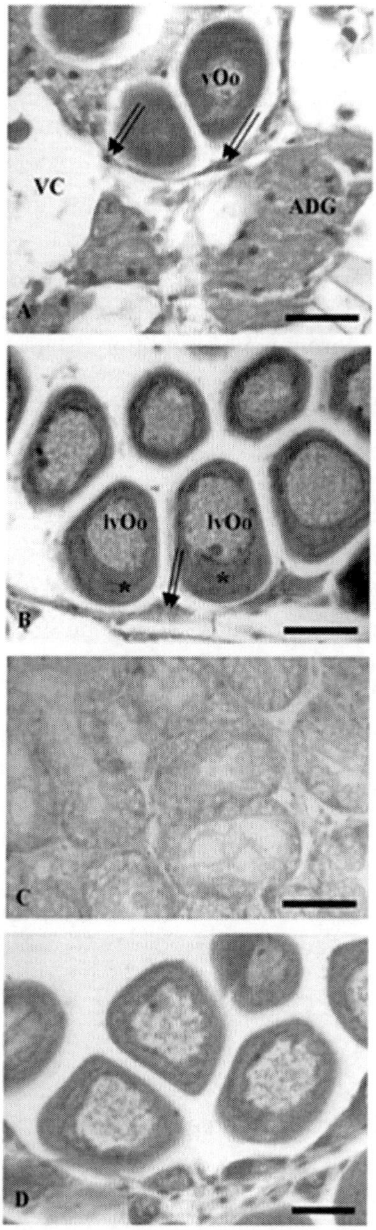

Figure 5. Immunolocalization of VTG in the ovary (A, B) and the digestive gland (C) of *Mytilus galloprovincialis*. The immuohistochemical signal, that appears as brown areas, is present in vitellogenic (vOo) (A) and full grown (lvOo) (B) oocytes, where the signal is located particularly in the juxta-nuclear area (B, asterisks). The immunohistochemical signal is present also in follicle (A, B, double arrows) and adipogranular cells (ADG) (A, B), while no signal was evident within vesicular cells (VC) (A). No signal was evident within digestive gland (C), as well as in the control sections performed omitting the primary antibody (D). Bars = 30 μm.

DISCUSSION

In this work we investigated the presence of VTG in the mussel Mytilus galloprovincialis, firstly by an indirect method (TEM) and then by direct ones, evaluating the VTG synthesis sites (RT-PCR and in situ hybridization) and the distribution of the protein (immunohistochemistry). We demonstrate the presence of both an autosynthesis (oocyte) and a heterosynthesis (follicle and adipogranular cells) taking place only in the gonad, while the digestive gland is not involved in VTG production.

Induction of the yolk protein precursor vitellogenin (Vtg) in plasma has proved to be a simple and sensitive biomarker for assessing exposure to environmental estrogens in fish (Tyler et al., 1998; Larsson, 1999) (xenoestrogens pollution), as it is synthesized under the estrogen control in mature females, although its synthesis can be induced in males and in immature females by specific environmental condition (Maradonna and Carnevali, 2007), and in mussels (Blaise et al., 1999; Riffeser and Hock, 2002; Marin and Matozzo, 2004; Porte et al., 2006). In recent times mussels have been largely used as sentinel species of the marine pollution, due to their features: indeed, these animals are sessile and filter-feeding organisms, show a wide distribution along most coastal zones in the world and have a high bioaccumulation and low biotransformation potential for both organic and inorganic contaminants (Ortiz-Zarragoitia and Cajaraville, 2006; Oehlmann and SchulteOehlmann 2002; Smolders et al. 2003).

The knowledge of VTG synthesis and its endocrine control in the mussels is scant and often based on indirect methods (Matsumoto et al., 2003; Osada et al 2003, 2004; Matsumoto et al., 2008). Particularly, we focused our attention on the mollusc present in the Gulf of Naples, the Mytilus galloprovincialis, in which TEM observations suggested that early vitellogenic oocytes, ovoid in shape, and vitellogenic oocytes, that are assuming a pear shape, are involved in an abundant protein synthesis, probably of VTG, as they are characterized by an extended synthesis apparatus constituted by numerous RER cisternae, extended Golgi complexes associated to nascent yolk platelets; also the cytoplasm of follicle cells connected to the oocyte and of adipogranular cells showed Golgi complex and RER cisternae. These observations were then corroborated by further analyses: indeed, firstly, RT-PCR showed a positive result just in the gonads containing oocytes in vitellogenesis; secondly, in situ hybridization showed that the expression of VTG in the oocytes begins early and stops when the oocytes are ready to be released into the lumen of the ovarian follicle and showed that also follicle and adipogranular cells are involved in VTG synthesis; finally, immunohistochemistry demonstrated the presence of VTG within germ, adipogranular and follicle cells, but not within digestive cells. Taken together, these results demonstrated that in Mytilus galloprovincialis both an autosynthesis, that takes place in the oocyte, and an heterosynthesis, that involves follicle and adipogranular cells, coexist. The contribute of follicle cells to VTG synthesis is in agreement with the results reported in Patinopecten (Osada et al., 2003, 2004) and in Haliotis (Matsumoto et al., 2008), although in our opinion the contribute to VTG synthesis by follicle cells is not much important, as the transfer of VTG from follicle cells to oocyte occurs mainly when the number of follicle cells is very limited.

Differently, an important contribution to VTG synthesis comes from adipogranular cells, as reported in Crassostrea virginica (Eckelbarger and Davis, 1996); indeed, these cells show a strong positivity to both mRNA and protein. However, the mechanism of VTG transfer to oocytes is still unknown: we hypothesize that VTG, once synthesized, is transferred to the growing oocytes via the hemolymph. This hypothesis is supported by previous studies (Pampanin et al., 2005; Ricciardi et al., 2008) reporting the presence of VTG into hemolymph, although by indirect methods. The finding of VTG into the hemolymph points out the possible use of this protein as a useful biomarker, due to the easiness of taking samples of hemolymph in our experimental model. The presence of VTG in the hemolymph confirms the presence of an heterosynthesis, that in Mytilus galloprovincialis does not seem to involve the digestive gland, where we obtained negative results from RT-PCR, ISH and immunohistochemistry.

In conclusion, our data are first records demonstrating that in Mytilus galloprovincialis VTG is synthesized inside gonad directly by the oocyte (autosynthesis) and by the follicle and adipogranular cells (heterosynthesis); particularly, the contribution of adipogranular cells could determine the presence of VTG in the hemolymph; in this way, the simple monitoring of this protein in the hemolymph could make the VTG a suitable and useful biomarker of xeno-estrogen marine pollution in an ideal candidate as sentinel species, like Mytilus galloprovincialis, where the concentration of VTG-like proteins has been reported to depend on estrogens.

ACKNOWLEDGMENTS

The Authors are grateful to John Wiley and Sons Inc for publication permission of some images already published in Agnese et al., 2013, J Cell Physiol 228 (3): 547-555; the copyright of the reused material is property of John Wiley and Sons Inc.

REFERENCES

Aarab, N., Minier, C., Lemaire, S., Unruh, E., Hansen, P.D., Larsen, B.K., Andersen, O.K. and Narbonne, J.F. (2004). Biochemical and histological responses in mussel (*Mytilus edulis*) exposed to North Sea oil and to a mixture of North Sea oil and alkylphenols. *Mar. Environ. Res*. 58: 437–441.

Aarab, N., Lemaire-Gony, S., Unruh, E. and Hansen, P.D. (2006). Preliminary study of responses in mussel (*Mytilus edilus*) exposed to bisphenol A, diallyl phthalate and tetrabromodiphenyl ether. *Aquatic Toxicology* 78: 86–92.

Agnese, M., Verderame, M., De Meo, E., Prisco, M., Rosati, L., Limatola, E., del Gaudio, R., Aceto, S. and Andreuccetti P. (2012). A network system for vitellogenin synthesis in the mussel *Mytilus galloprovincialis. J. Cell Physiol.* doi: 10.1002/jcp.24161

Blaise, C., Gagné, F., Pellerin, J. and Hansen, P.D. (1999). Determination of vitellogenin-like properties in *Mya arenaria* hemolymph (Saguenay Fjord, Canada): a potential biomarker for endocrine disruption. *Environ. Toxicol*. 14:455–65.

De Jong-Brink, M., Boer, H.H. and Joosse, J. (1983). Mollusca. In: Adiyodi KG, Adiyodi RG, editors. *Reproductive Biology of Invertebrates* Vol. 1. Wiley Chichester p 297–335.

Duinker, A., Haland, L., Hovgaard, P. and Mortensen, S. (2008). Gonad development and spawning in one and two year old mussels (*Mytilus edulis*) from Western Norway. *J. Mar. Biol. Ass U.K.* 88(7): 1465–1473.

Eckelbarger, K.J. and Davis, C.V. (1996). Ultrastructure of the gonad and gametogenesis in the eastern oyster, *Crassostrea virginica*. Ovary and oogenesis. *Marine Biology* 127: 79–87.

Gagné, F., Blaise, C., Salazar, M., Salazar, S. and Hansen, P.D. (2001). Evaluation of estrogenic effects of municipal effluents to the freshwater mussel *Elliptio complanata*. *Comp. Biochem. Physiol* 128C: 213–255.

Larsson, D.G.J., Adolfsson-Erici, M., Parkkonen, J., Pettersson, M., Berg, A.H., Olsson, P.E. and Forlin, L. (1999). Ethinyloestradiol - an undesired fish contraceptive? *Aquatic Toxicol.* 45:91-97

Maradonna, F. and Carnevali, O. (2007). Vitellogenin, zona radiata protein, cathepsin D and heat shock protein 70 as biomarkers of exposure to xenobiotics. *Biomarkers* 12, 240-255.

Marin, M.G. and Matozzo, V. (2004). Vitellogenin induction as a biomarker of exposure to estrogenic compounds in aquatic environments. *Marine Poll Bull* 48: 835-839.

Matozzo, V. and Marin, M.G. (2005). Can 4-nonylphenol induce vitellogenin-like proteins in the clam *Tapes philippinarum*? *Environ. Res.* 97: 43-49.

Matozzo, V., Gagnè, F., Marin, M.G., Ricciardi, F. and Blaise, C. (2008). Vitellogenin as biomarker of exposure to estrogenic compounds in acquatic invertebrates: A review. *Environ. Int* 34: 531-545.

Matsumoto, T., Nakamura, A.M., Mori, K. and Kayano, T. (2003). Molecular characterization of a cDNA encoding putative vitellogenin from the Pacific oyster *Crassostrea gigas*. *Zool. Sci.* 20: 37–42.

Matsumoto, T., Yamano, K., Kitamura, M. and Hara, A. (2008). Ovarian follicle cells are the site of vitellogenin synthesis in the Pacific abalone *Haliotis discus hannai*. Comp Bioch and Physiol- Part A: *Molecular and Integrative Physiology* 149:293-298.

Mazzi, V. (1977). Manuale di tecniche istologiche e istochimiche. In Mazzi V., Editor Padova (Italy). p 63, 278.

Ortiz-Zarragoitia, M. and Cajaraville, M.P. (2006). Biomarkers of exposure and reproduction-related effects in mussels exposed to endocrine disruptors. *Arch. Environ. Contam. Toxicol*. 50: 361–369.

Oehlmann, J. and Schulte-Oehlmann, U. (2002) Molluscs as bioindicators. In: Markert BA, Breure AM, Zechmeister HG (eds.) Bioindicators and biomonitorings. *Elsevier*, New York, NY, p 577.

Osada, M., Takamura, T., Sato, H. and Mori, K. (2003). Vitellogenin synthesis in the ovary of the scallop, *Patinopecten yessoensis*: control by estradiol-17b and the central nervous system. *J. of Exp. Zool.* 299A: 172–179.

Osada, M., Harata, M., Kishida, M. and Kijima, A. (2004). Molecular cloning and expression analysis of vitellogenin in scallop, *Patinopecten yessoensis* (Bivalvia, Mollusca). *Mol. Reprod. Dev*. 67: 273-281.

Pampanin, D.M., Marangon, I., Volpato, E., Campesan, G. and Nasci, C. (2005). Stress biomarkers and alkali-labile phosphate level in mussels (*Mytilus galloprovincialis*) collected in the urban area of Venice (Venice Lagoon, Italy). *Environ. Pollut* 136:103–7.

Pipe, R.K. (1985). Seasonal cycles in and effects of starvation egg development in *Mytilus edulis*. *Ecol. Prog. Ser.* 24: 121-125.

Polzonetti-Magni, A.M., Mosconi, G., Soverchia, L., Kikuyama, S. and Carnevali, O. (2004). Multihormonal control of vitellogenesis in lower vertebrates. In *International Review of Cytology* 239: 1-46.

Porte, C., Janer, G., Lorusso, L.C., Ortiz-Zarragoitia, M., Cajaraville, M.P., Fossi, M.C. and Canesi, L. (2006). Endocrine disruptors in marine organisms: Approaches and perspectives. *Comp. Biochem. and Physiol. Part C* 143: 303–315.

Prisco, M., Valiante, S., Romano, M., Ricchiari, L., Liguoro, A., Laforgia, V., Limatola, E. and Andreuccetti, P. (2004). Ovarian follicle cells in *Torpedo marmorata* synthesize vitellogenin. *Mol. Reprod. Dev.* 67:424-429.

Puinean, A.M., Labadie, P., Hill, E.M., Osada, M., Kishida, M., Nakao, R., Novillo, A., Callard, I.P. and Rotchell, J.M. (2006). Laboratory exposure to 17β-estradiol fails to induce vitellogenin and estrogen receptor gene expression in the marine invertebrate *Mytilus edulis*. *Aquat Toxicol* 79:376–83.

Quinn, B., Gagné, F., Costello, M., Mackenzie, C., Wilson, J. and Mothersill, C. (2004). The endocrine disrupting effect of municipal effluent on the zebra mussel (*Dreissena polymorpha*). *Aquat. Toxicol.* 66: 279–292.

Ricciardi, F., Matozzo, V. and Marin, M.G. (2008). Effects of 4-nonylphenol exposure in mussels (*Mytilus galloprovincialis*) and crabs (*Carcinus aestuarii*) with particular emphasis on vitellogenin induction. *Marine Poll Bull* 57: 365–372.

Riffeser, M. and Hock, B. (2002). Vitellogenin levels in mussel hemolymph- a suitable biomarker for the exposure to estrogens? *Comp. Biochem. Physiol. C Toxicol. Pharmacol.* 132(1):75-84.

Smolders, R., Bervoets, L., Wepener, V. and Blust, R. (2003) A conceptual framework for using mussels as biomonitors in whole effluent toxicity. *Human Ecol Risk Assess* 9:741–760.

Tyler, C.R., Jobling, S. and Sumpter, J.P. (1998). Endocrine disruption in wildlife: A critical review of the evidence. *Crit. Rev. Toxicol.* 28:319-361.

Verslycke, T., Vandenbergh, G.F., Versonnen, B., Arijs, K. and Janssen, C.R. (2002). Induction of vitellogenesis in 17-ethylestradiol-exposed rainbow trout (*Oncorhynchus mykiss*): a method comparison. *Comp. Biochem. Physiol.* 132:483– 492.

Versonnen, B.J., Goemans, G., Belpaire, C. and Janssen, C.R. (2004). Vitellogenin content in European eel (*Anguilla anguilla*) in Flanders, Belgium. *Environ. Poll* 128: 363–371.

Wallace, R.A. (1985). Vitellogenesis and oocyte growth in nonmammalian vertebrates. In: Browder L. editor, Development Biology 1: Oogenesis. University of Calgary, Calgary, Alberta, Canada. Vol 3, p 127-177.

Zorita, I., Ortiz-Zarragoitia, M., Soto, M. and Cajaraville, M.P. (2006). Biomarkers in mussels from a copper site gradient (Visnes, Norway): an integrated biochemical, histochemical and histological study. *Aquat. Toxicol.* 78(1): 109–16.

In: Mussels: Ecology, Life Habits and Control
Editors: Jarek Nowak and Michal Kozlowski
ISBN: 978-1-62618-083-3

Chapter 6

BRACHIDONTES SOLISIANUS (BIVALVIA: MYTILIDAE) RE-COLONIZATION DYNAMICS ON A SANDSTONE REEF AT PERNAMBUCO - BRAZIL

Mariana Alves de Guimaraens[1*], Jessica Pires Sá Carneiro[1], Betty Rose de Araújo Luz[1] and Janilson Felix da Silva[2]
[1]Universidade de Pernambuco; Rua Arnóbio Marques, Recife - PE, Brazil
[2]Universidade Federal de Pernambuco; Rua Professor Moraes Rego, Recife – PE, Brazil

ABSTRACT

Three sub-environments were identified on the reef flat at Piedade Beach. Each sub-environment was characterized by different organisms on the intertidal zone. The mesointertidal contained the algal sub-environment, while the mussel *Brachidontes solisianus* (d´Orbigny, 1846) dominated the bivalve sub-environment, and patches of Cirripedia dominated the higher intertidal. Mussels were collected by scraping, three quadrats in each sub-environment each month. The mussels were classified in three size classes according to shell length. Each month´s sample was counted and measured to evaluate re-colonization patterns and correlations with environmental factors, such as air temperature, rainfall and average algal biomass, within distinct sub environments between March and December of 2007. The field data show a positive relationship between size classes in the *Brachidontes solisianus* colonization experiments in the Cirripedia, mussel and algal sub-environments. These results suggest that mussel recruitment tends to occur closer to older individuals, although the field observations indicated that the patch closure was the principal process of recovery from disturbance. A positive correlation was found between rainfall and mussel recruitment in the Cirripedia sub-environment, and a positive relationship between bivalve recruitment and macroalgal biomass was found in the *Brachidontes* and algal sub environments. A covariance analysis showed significant differences in bivalve density among the three sub-

* Corresponding Author: mguimaraens@hotmail.com, mariana.guimaraens@pq.cnpq.br.

environments. *Brachidontes* recruitment was highest in October for the algal sub-environment (5483 individuals/m^2), whereas colonization was highest in June 2007 in the invertebrates sub- environments.

Keywords: Mussel bed, macroalgae biomass, sandstone reefs, *Brachidontes* recruitment

INTRODUCTION

Mussels in intertidal communities are dominant organisms on rocky coasts and reefs because they are often superior competitors for space on these communities (Navarrete and Castilla, 1990; Barbiero et al., 2011; Santos and Boehs, 2011) where desiccation is reduced. Borthagaray and Carranza (2007) showed that mussels can add complexity to benthic environments and increase the richness of species and the diversity in the intertidal zone. *Brachidontes solisianus* is a small species of bivalve belonging to the family Mytilidae. The free swimming larvae of this species settle on shallow hard substrate and basal algal surfaces attached by their byssus (Monteiro-Ribas et al., 2006). These aggregations may decrease mortality resulting from desiccation and temperature stress. *Brachidontes solisianus* populations are present from Mexico to Uruguay on the western Atlantic (Rios, 1994). Patrício et al. (2006) conducted recolonization experiments showing that bivalves are one of the principal organism component of the rocky coast community recolonization. However, Barbiero et al. (2011) only found *Brachidontes solisianus* in control treatments in a study of ecological succession in a rocky intertidal zone community of south-eastern Brazil. Bottom-up effects on community structure have been shown to control algal growth and to mediate competition with invertebrates in reefs and rocky coast communities of eutrophized and overfished areas (Coutinho et al, 1993; Lapointe 1997; Vinueza et al., 2006; Sotka and Hay, 2009). Similarly Figueiredo et al. (2008) found phase shifts on Brazilian reefs. As a result from these phase shifts animal-dominated communities were replaced by algae-dominated communities. Porri et al. (2006) showed that an interaction between time and location influences the settlement of the bivalve *Perna perna* (Linnaeus, 1758), which suggested that hydrodynamics can affect re-colonization patterns, and emphasized the significance of bottom-up effects, such as recruitment, for population structure. Moreover, Nolesso et al. (2008) suggested that the ephemeral alga *Ulva* spp. prevented oyster recruitment on collecting structures and that temperature and salinity were positively correlated with spat recruitment.

Brachidontes solisianus occurs in higher salinities than *Brachidontes darwinianus* (d´Orbigny, 1846), and is found higher on the shore when the two species co-occur (Tanaka, 2005). *B. solisianus* generally recruit in the intertidal zone on hard rocky surfaces or on the basal regions of algae through their byssus (Monteiro-Ribas et al., 2006). Petersen et al. (1986), recorded the gradual decline on the population of *Brachidontes solisianus* on a coast at São Sebastião – SP. Suggesting that temperature and desiccation had negative effects on the on the bivalves and that competition with barnacles also occurred. Therefore, these authors concluded that the dominance of mussels on intertidal zones of rocky coasts and reefs is typical within temperate climates and temporary in tropical environments. These studies illustrate the linkage between recruitment and population dynamics. Indeed recruitment rates for marine invertebrates can be used in ecological models to evaluate population stability

(Guimaraens et al., 2007; Guimaraens et al., 2011a). In view of considering the patchy distribution of the bivalve population in the intertidal zone of Piedade Reef and the seasonality of the reef community (Costa, 2007; Guimaraens et al., 2011b; Guimaraens et al., 2011b), the objective of this research was to evaluate the temporal and spatial variability of *Brachidontes solisianus* re-colonization in a mussel bed at the reef located in Pernambuco, Brazil. The study tested the hypothesis that recruitment and size classes in bivalves are related to environmental factors such as air temperature, rainfall and algal biomass.

METHODS

Study Site

The study site is located at Piedade Beach, Jaboatão dos Guararapes district, on the southern coast of Pernambuco State. The coordinates of the site are 08° 11´03.45´´ and 08° 11´08.48´´ S, and 34° 55 03.24´´ and 34° 55´ 04.66´´ W (Simões et al,. 2009); where; the sandstone reef area is approximately 234 km^2. Samples were taken from a *Brachidontes solisianus* bed on the south end of the reef´s flat intertidal zone, where algal growth was more abundant closer to the sea. In contrast, Cirripedia (*Chthamalus bisinuatus* Pilsbry, 1916 and *Tetraclita stalactifera* Lamarck, 1818) patches were primarily observed mainly farther from the water (Guimaraens et al., 2011a). In addition to the occurrence of filamentous and foliose green algae such as *Ulva* spp. and turf-forming algae, the mollusks *Nodlitorina, Collisella, Fissurella* and *Stramonita* occurred frequently in the intertidal zone of the reef (Costa, 2007). The environmental data for March through December 2007 used in the statistical analyses were provided by the state government climatology station. During the study period, the minimum average air temperature occurred in July (24.33°C) and the maximum in December (26.96°C). The rainfall reached its maximum in June (325.9 mm). The minimum rainfall occurred in October when no rainfall was recorded for Jaboatão dos Guararapes District (Table 1).

Table 1. Environmental data from the study area and algal biomass for each sub-environment in grams of dry weight/m^2

Months	Average Air Temperature (°C)	Rainfall (mm)	Cirripedia	*Brachidontes*	Algal Biomass
April	26.68	227.7	0	0	13.91
May	25.77	201.9	0	0	16.49
June	24.97	325.9	8.67	7.67	12.41
July	24.33	208.4	3.33	8.67	43.98
August	24.59	176.7	46.60	3	106.37
September	25.05	116.4	117	0	90.80
October	25.96	0	29	0	258.81
November	26.56	0.1	43	0	56.73
December	26.96	30	1.67	0	32.74

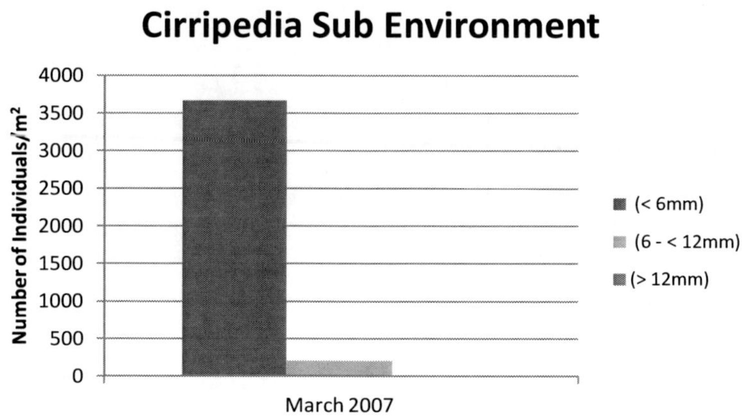

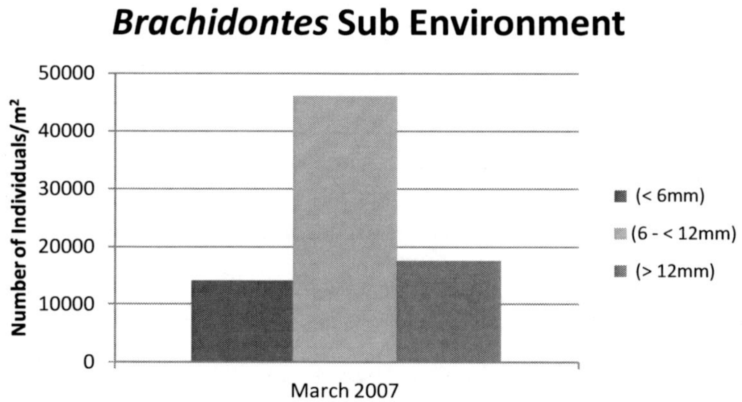

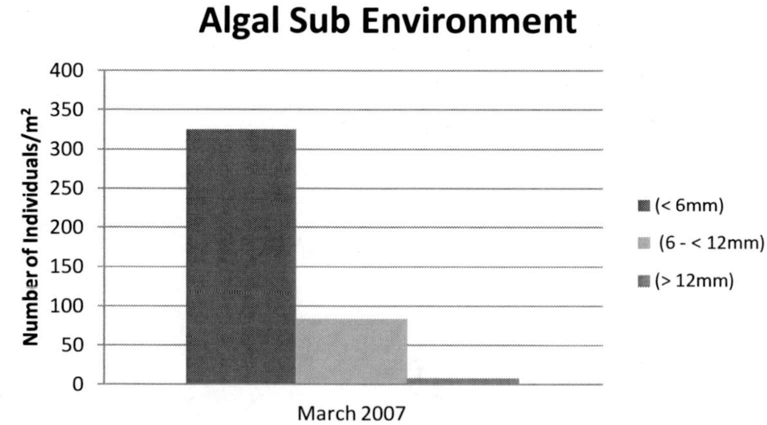

Figure 1. *Brachidontes solisianus* density per m^2 collected on March, 2007, on non-manipulated plots in the three sub-environments.

The rainy season extends from March to September with a monthly rainfall greater than 100 mm and the dry season extends from October to February with rainfall less than 100 mm (Cavalcanti and Kempf, 1969). The reef sampled in this study is located approximately 3 kilometers north of the Jaboatão River estuary, which has a history of poor water quality. Sites close to the shore were classified as mesotrophic and eutrophic at Piedade (Passavante and Feitosa, 1995), since planktonic organic production may have decreased during the rainy season due to the influence of suspended solids in this area.

Field Sampling

Samples were collected from 3 sub environments, whose locations were related to the zonation of the *Brachidontes* bed (Cirripedia in the higher intertidal, *Brachidontes* and algal sub-environments in the meso-intertidal zone), to evaluate recruitment and patch closure associated with lateral migration of bivalves. Three random quadrats were selected and scraped in each sub environment along 1-m-long transects parallel to the coast (Cirripedia and *Brachidontes*) and along 2-m-long transect in the algal sub-environment. Transects were selected to allow us to sample a uniform community distribution in view of the patchy structure of the reef heterogeneity of the substrate and the different bivalve densities in each sub-environment. The first sample, collected in March 2007, was used to characterize the spatial distribution of the different bivalve size classes on the reef (Figure1). Thereafter, the quadrats were scraped monthly from April to December 2007 to evaluate re-colonization. The quadrat size used to sample the Cirripedia and *Brachidontes* sub environments (Guimaraens and Coutinho, 1996; Skinner and Coutinho, 2005) was 10 x 10 cm, whereas quadrat size in the algal sub environment was 20 x 20 cm to facilitate the total counting and individual measurement to characterize the *Brachidontes solisianus* population (Figure 1).The quadrat size and number was based on previous work with mussels and algae at Piedade Reef (Almeida et al., 2008; Guimaraens et al., 2011b). The samples were taken to the Marine Biology Laboratory at Universidade de Pernambuco and frozen for further sorting.

Laboratory Methods and Statistical Analyses

To separate the macroalgae from the macrofauna, the samples were washed with flowing water over a 300-micrometer mesh seize. The sorted fauna was fixed with 70% alcohol. The macroalgae were frozen for further dry weight measurements. After defrosting, the algae were placed in pre-weighted aluminum baskets and dried at 60°C for 24 hours; the samples were then weighted on a high-precision scale. *Brachidontes solisianus* individuals from each sampled quadrat were measured and counted. According to Rios (1994), this species reaches the adult stage at a shell length between 12 and 17 mm. Accordingly, three size classes based on shell length: were defined to identify the developmental stage of the bivalve: recruits, with a shell length up to 6 mm, immature individuals, from 6 to less than 12 mm; and adult individuals, with a shell length of 12 mm and grater.

To evaluate differences between the average re-colonization densities, we conducted a repeated measurements ANOVAs for each sub-environment with JMP software (SAS Institute). The data analysis were obtained by adding the densities of given size classes for each sampled plot. This approach was valid because the sampled recruitment quadrats were the same. A Neuman-Keuls a posteriori test was used to detect differences between the monthly average bivalve densities. The density data (x) were transformed by the square root of x +0.5 to meet homoscedasticity assumptions. In addition the total monthly density values of the several size classes of bivalves and of algal biomass, measured from April through December 2007, were used to calculate Pearson correlation coefficients between the density values rainfall and average air temperature data, JMP software was used for these calculations. To compare the density of recruits and immature individuals among the populations of the three sub- environments, an analysis of covariance was performed with the Biostat software.

RESULTS AND DISCUSSION

The data collected in March, 2007, showed a greater abundance of recruits in the *Brachidontes* sub-environment (14099 individuals/m^2) than in the Cirripedia (3666 individuals/ m^2) and algal (324 individuals/m^2) sub-environments (Figure 1). The highest density of immature individuals (46129 individuals/m^2) was also recorded in the *Brachidontes* sub-environment. In the first sample, no adult individuals were found in the Cirripedia sub-environment, and individuals with a shell length of 12 mm and longer were more abundant in the *Brachidontes* sub-environment (17565 individuals/m^2). During the artificial disturbance experiment, which was performed by scraping quadrats in the sub-environments, the most abundant individuals in every sampled belonged to the recruit size class (Figure 2). Within the algal sub-environment, the greatest density of recruits occurred in October, 2007, following the period of the highest ephemeral algal abundance on the reef. This event corresponded to a peak algal recruitment on the reef of nearly 250 g dry wt/m^2 (Figure 2, Simões et al. 2009, Guimaraens et al., 2011b). In contrast the highest recruitment in the *Brachidontes* sub environment occurred in June and July, 2007. In the Cirripedia sub-environment, *Brachidontes solisianus* recruitment reached its peak in June, 2007, before the ephemeral algal bloom occurred in the reef at the end of the rainy season. However, no statistical differences in monthly invertebrate recruitment values were found by the analyses of variance in the *Brachidontes* sub environment ($p > 0.05$). Moreover seasonal re-colonization dynamics, showed significantly different patterns in the Cirripedia and algal sub-environments ($p < 0.05$). Although immature individuals and adults were less abundant than smaller individuals in the *Brachidontes* sub-environment, these size classes showed a patch occupation pattern similar to the pattern of recruits (Figure 2) in the *Brachidontes* and algal sub-environments.

Brachidontes solisianus recruitment was positively correlated with immature individuals and rainfall in the Cirripedia sub-environment samples (Table 2). The absence of adults in these samples suggests a limitation of the vertical distribution of the studied species by desiccation stress. In the *Brachidontes* sub-environment samples, bivalve recruits were positively correlated with immature and adult individuals and the three size classes were positively correlated with algal biomass (Table 3). A negative correlation also occurred between temperature and the three size classes. In the algal sub-environment, there was a positive correlation between bivalve recruitment and immature and adult individuals (Table 4). The density of the different size classes was also correlated with algal abundance. Algal biomass was negatively correlated with temperature in the mussel sub-environment.

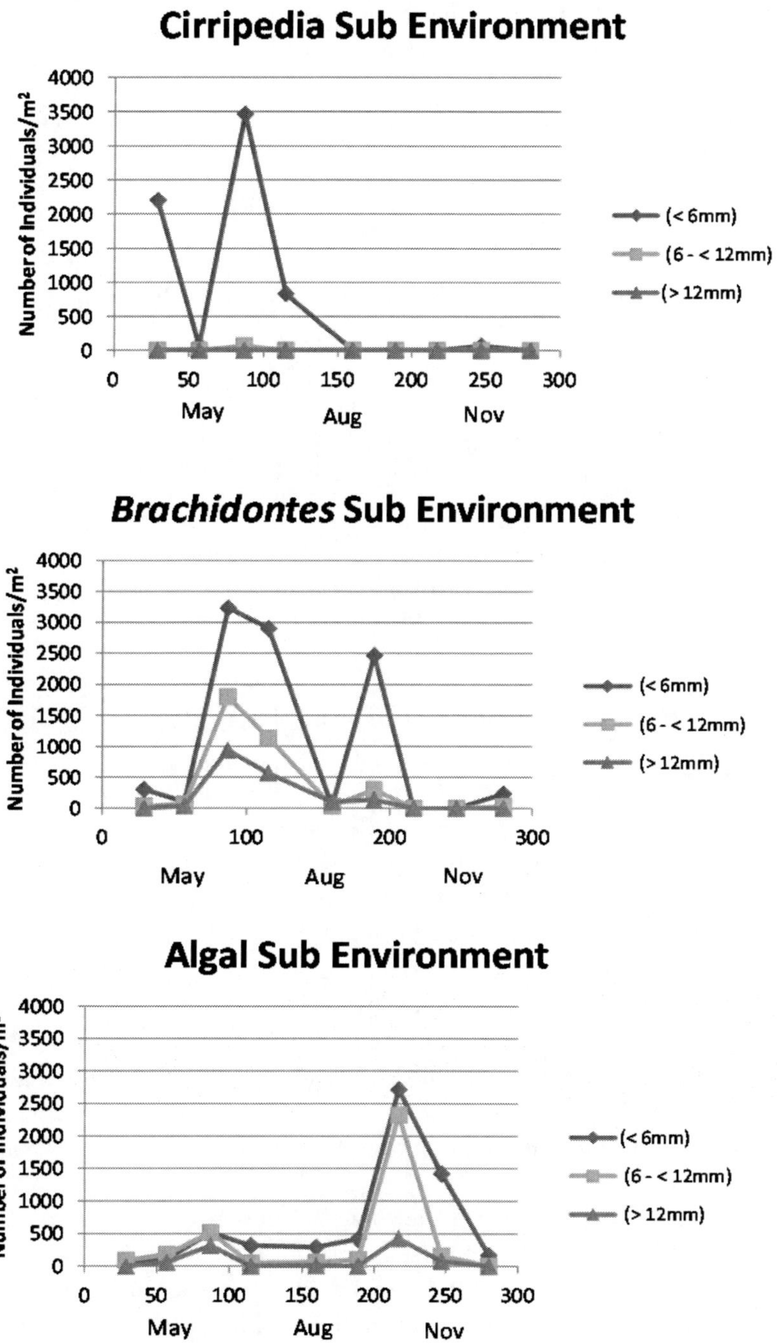

Figure 2. *Brachidontes solisianus* density per m^2 collected in the re-colonization experiment from April to December, 2007, in the three sub-environments.

The correlations among the three bivalve size classes in terms of patch occupation suggest that recruits settle next to older individuals of the same species (Rodríguez et al., 1993). Recruitment was also positively correlated with algae biomass, such as those from the genus *Ulva* (Simões et al., 2009, Guimarens et al., 2011b).

Table 2. One – way ANOVA for monthly bivalve density variation in the Cirripedia sub-environment

Source of Variation	d.f.	SS	MS	F	p-level	F crit
Between Groups	8	68.81803	8.6023	2.7465	0.0357	2.5102
Within Groups	18	56.377384	3.1321			
Total	26	125.1954				

Table 3. One – way ANOVA for monthly bivalve density variation in the *Brachidontes* sub-environment

Source of Variation	d.f.	SS	MS	F	p-level	F crit
Between Groups	8	146.339	18.2924	2.4834	0.052	2.5102
Within Groups	18	132.5829	7.3657			
Total	26	278.922				

Table 4. One-way ANOVA for monthly bivalve density variation in the Algal sub-environment

Source of Variation	d.f.	SS	MS	F	p-level	F
Between Groups	8	337.8753	42.2344	4.492805	0.0039	2.5102
Within Groups	18	169.2082	9.4005			
Total	26	507.0834				

In addition, the growth of ephemeral algae has been shown to be positively correlated with *Brachidontes solisianus* succession (Costa, 2007), and bivalves were found to settle after algal colonization in field experiments at Portugal intertidal coasts as well (Patrício et al., 2006). However, the growth of invasive green algae has been shown to limit the performance of bivalve recruits in a soft-bottom benthic environment in Australia (Gribben et al., 2009). A regression analysis showed a significant relationship between recruits and immature individuals ($p<0.05$) for the three sampling sites. However, a covariance analysis suggested that the distribution of the invertebrate densities among the three sub-environments were statistically different. We found higher number of recruits than immatures and adults in the *Brachidontes* sub- environment.

However, succession experiments in the *Brachidontes* beds showed a different pattern, visually, lateral migration appeared to be responsible for most occurrences of patch closure in small gaps (Tanaka and Magalhães, 2002; Tanaka, 2005; Costa, 2007). These patch closure events were followed by the growth of the population of Cirripedia (Costa, 2007; Guimaraens et al., 2011a), during the dry season on Piedade Reef. The Cirripedia can recruit to *Brachidontes solisianus* shells (Guimaraens et al., 2011a). In addition, the invasive species *Brachidontes pharaonis* (Fischer, 1870) shows a high plasticity of physiological rates over a wide range of salinities and temperatures and has the potential to compete with indigenous bivalves in the Mediterranean (Sarà et al., 2008). This species shows delayed colonization of rocky platforms along the Israeli coast. This outcome is related to reduced sediment deposition, lower mortality rates and interespecific competition with native mussel species (Rilov et al., 2004).

Table 5. Correlation coefficients for *Brachidontes solisianus* re-colonization and environmental parameters in the Cirripedia sub-environment

	Recruits	Immatures	Temperature	Rainfall
Immatures	**0.8133**			
Temperature	-0.1174	-0.2659		
Rainfall	**0.7577**	0.6019	-0.5229	
Algal Biomass	-0.3799	-0.1839	-0.2625	-0.2969

The correlations in bold are statistically significant (p<0.05).

Table 6. Correlation coefficients for *Brachidontes* re-colonization and environmental parameters in the *Brachidontes solisianus* sub-environment

	Recruits	Immatures	Adults	Temperature	Rainfall
Immatures	**0.8797**				
Adults	**0.8689**	**0.9996**			
Temperature	**-0.8632**	**-0.7372**	**-0.7314**		
Rainfall	0.6236	**0.7195**	0.7161	-0.5273	
Algal Biomass	**0.8153**	**0.9331**	**0.9296**	**-0.7593**	0.6345

The correlations in bold are statistically significant (p<0.05).

Table 7. Correlation coefficients for *Brachiodents solisiams* re-colonization and environmental parameters in the Algal sub-environment

	Recruits	Immatures	Adults	Temperature	Rainfall
Immatures	**0.8948**				
Adults	**0.9066**	**0.8662**			
Temperature	0.1610	0.1214	0.2272		
Rainfall	-0.6126	-0.4561	-0.4791	-0.5229	
Algal Biomass	**0.8425**	**0.8931**	**0.7023**	-0.0732	-0.5659

The correlations in bold are statistically significant (p<0.05).

Likewise, recruitment experiments at Recife Harbor on submersed non-natural surfaces showed a decline in barnacle recruitment, related to rainfall and low salinity, followed by an increase in exotic mussel colonization (Nery et al., 2008). Analysis of the mussel recruitment data showed statistically significant ($p < 0.05$) recolonization with a peak in October 2007 in the algal sub environment.

This finding can be related to high recruitment in ephemeral algal biomass (Guimaraens et al., 2011b) and less salinity stress. The lack of statistical significance for the ANOVAs of the bivalve recruitment data from the *Brachidontes solisianus* sub-environment appears to result from the patchy spatial distribution of mussel and from impacts by beach visitors. Although the re-colonization of experimental gaps occurred during the field study environmental stressors such as desiccation, interspecific competition, disturbance by predation and trampling may limit the development of mussel beds at Piedade Reef.

Conclusion

The *Brachidontes solisianus* population showed different re-colonization patterns in the different sub-environments according to the characteristics of these micro-habitats. The difference in these patterns was illustrated by the colonization of the Cirripedia sub-environment, located in the higher intertidal, particularly during the rainy season. In contrast, various recolonization events occurred during the year in the *Brachidontes* and algal sub-environments. In the algal sub-environment, a peak in *Brachidontes solisianus* density was recorded in October 2007. This peak was associated with the recruitment of ephemeral algae. Two modes of space occupation by the bivalve were observed: recruitment and lateral migration. Mussel colonization was significantly correlated with algal biomass in the *Brachidontes* and algal sub environments. The correlation among bivalve recruits, immature and adult individuals suggests that recruitment was associated with the presence of older individuals. Therefore, the re-colonization experiment conducted during this study suggests that the studied community has the potential to facilitate the recovery of the mussel population and to maintain the population structure of the mussels in the intertidal of Piedade Reef.

Acknowledgments

This study received financial support from CNPq-Facepe through "Primeiros Projetos" Grant 2006-2008 (APQ.0129.2.05/06).

References

Almeida, L. R.; Guimaraens, M. A.; Luz, B. R. A. and Porto-Neto, F. F. 2008. Interação entre Processos de Colonização e Sucessão Ecológica de Comunidades Bioincrustantes na Praia de Piedade. *Tropical Oceanography*, 36: 40-46.

Barbiero, D. C.; Macedo, B. P.; Masi, B. P. and Zalmon, I. R. 2011. A removal disturbance effect on the ecological succession of a benthic intertidal community south-eastern Brazil. *Marine Biodiversity Records*, 4: 1-8.

Borthagaray, A.I. and Carranza, A. 2007. Mussels as ecosystems engineers: their contribution to species richness on a rocky littoral community. *Acta Oecologica*, 31: 243-454.

Cavalcanti, L. B. and Kempf, M. 1969. Estudo da plataforma continental na área do Recife (Brasil). II. Meteorologia e Hidrologia. *Trabalhos Oceanográficos*, 9: 149-158.

Costa, D. C. P. 2007. Estudo da sucessão e do recrutamento de invertebrados e cobertura de algas na zona entremarés do recife arenítico da Praia de Piedade, Jaboatão - PE. B.S. Dissertation, Universidade de Pernambuco, 49 p.

Coutinho, C.; Magalhães, C.; Villaça, R.C.; Guimaraens, M.A.; Silva Jr., M. A. and Muricy, G. 1993. Influência antrópica sobre os ecossistemas coralíneos de Abrolhos. *Acta biológica Leopoldensia*, 27: 133-144.

Figueiredo, M. A. O.; Horta, P.; Pedrini A. G. and Nunes J. M. C. 2008. Benthic algae of the coral reefs of Brazil: a literature review. *Oecologia brasiliensis*, 12: 258-269.

Gribben, P.E.; Wright, J.T.; O'Connor, W.A.; Doblin, M .A.; Eyere, B. and Steinberg, P.D. 2009. Reduced performance of native infauna following recruitment to habitat-forming invasive marine alga. *Oecologia*, 158: 733-745.

Guimaraens, M.A. and Coutinho R. 1996. Spatial and temporal variation of benthic marine algae at the Cabo Frio upwelling region, Rio de Janeiro, Brazil. *Aquatic Botany*, 52: 283–299.

Guimaraens, M. A.; Costa, D. C. P.; Costa, M. I. S.; Souza, J. R. B. and Alves, A. M. F. 2007. Simulações da dinâmica, regulação populacional e recrutamento de cirrípedes e bivalves marinhos bentônicos. *Tropical Oceanography*, 35: 17-31.

Guimaraens, M. A.; Luz, B. R. A.; Silva, J. F. and Carneiro, J. P. S. 2011a. Modeling the succession of barnacles and mussels on a sandstone reef in Pernambuco State, Brazil. *Hydrobiologia*, 658: 365-372.

Guimaraens, M. A.; Silva, J. A. M. and Falcão, D., 2011b. Temporal variability in macroalgae recruitment and succession on sandstone reefs at Piedade Beach – PE, Brazil. *Neotropical Biology and Conservation*, 6: 170-177.

Lapointe, B.E. 1997. Nutrient thresholds for bottom-up control of macroalgal blooms on coral reefs in Jamaica and southeast Florida. *Limnology and Oceanography*, 42: 1119-1131.

Monteiro-Ribas, W.; Rocha-Miranda, F.; Romano, R. C. and Quintanilha, J. 2006. Larval development of *Brachidontes solisianus* (Bivalvia, Mytilidae), with notes on differences between its hinge system and those of the mollusk *Perna perna*. *Brazilian Journal of Biology*, 66: 109-116.

Navarrete, S. A. and Castilla, J. C. 1990. Barnacle walls as mediators of intertidal mussel recruitment: effects of patch size on the utilization of space. *Marine Ecology Progress Series*, 68: 113-119.

Nery, P. P. C. F.; Leitão, S. N.; Fernandes, M. L. B.; Silva, A. K .P. and Chaves, A. C. 2008. Recrutamento e sucessão ecológica da macrofauna incrustante em substratos no Porto do Recife – PE, Brasil. *Revista Brasileira de Engenharia de Pesca*, 3: 51-61.

Nolesso, R. C.; Paresque, K.; Piumbini, P. P.; Tonini; Almeida, L. G. and Nickel, V. M. 2008. Oyster spat recruitment in Espírito Santo state, Brazil, using recycled materials. *Brazilian Journal of Oceanography*, 56: 281-288.

Passavante, J.Z.O. and Feitosa, F.A.N. 1995. Produção primária do fitoplâncton da plataforma continental de Pernambuco (Brasil): Área de Piedade. *Boletim Técnico Científico CEPENE*, 3: 7-22.

Patrício, J.; Salas, F.; Pardal, M. A.; Jorgensen, S. E. and Marques, J. C. 2006. Ecological indicators performance during a re-colonization field experiment and its compliance with ecosystem theories. *Ecological Indicators*, 6: 43-57.

Petersen, J. A.; Sutherland, J. P. and Ortega, S. 1986. Patch dynamics of mussel beds near São Sebastião (São Paulo), Brazil. *Marine Biology*, 93: 389-393.

Porri, F.; McQuaid, C. D. and Radloff, S. 2006. Temporal scales of variation in settlement and recruitment of the mussel *Perna perna* (Linnaeus, 1758). *Journal of Experimental Marine Biology and Ecology*, 332: 178-187.

Rilov, G.; Benayahu, Y and Ganith, A. 2004. Prolonged lag in population outbreak of an invasive mussel: a shifting-habitat model. *Biological Ivasions*, 6: 347-364.

Rios, E. C. 1994. Seashells of Brazil. Rio Grande, Brazil, Museu Oceanográfico da Furg.

Rodríguez, S. R.; Ojeda, F. P. and Inestrosa, N. 1993. Settlement of benthic marine invertebrates. *Marine Ecology Progres Series*, 97: 193-207.

Santos, J. J. B and Boehs, G., 2011. Spatial-temporal distribution and recruitment of *Stramonita haemastoma* (Linnaeus, 1758) (Mollusca) on a sandstone bank in Ilhéus, Bahia, Brazil. *Brazilian Journal of Biology*, 4: 799-805.

Sarà, G.; Romano, C.; Widdows, J. and Staff, F. J. 2008. Effect of salinity and temperature on feeding physiology and scope for growth of an invasive species (*Brachidontes pharaonis* – Mollusca: Bivalvia) within the Mediterranean sea. *Journal of Experimental Marine Biology and Ecology,* 363: 130-136.

Simões, I. P.; Guimaraens, M. A.; Oliveira-Carvalho, M. F.; Valdevino, J. and Pereira, S. M. B. 2009. Avaliação florística e sucessão ecológica das macroalgas em recifes na praia de Piedade – PE. *Neotropical Biology and Conservation*, 4: 49-56.

Skinner, L.F. and Coutinho, R. 2005. Effect of microhabitat distribution and substrate roughness on barnacle *Tetraclita stalactifera* (Lamark, 1818) settlement. *Brazilian Archives of Biology and Technology*, 48: 109-113.

Sotka, E. E. and Hay, M. E. 2009. Effects of herbivores, nutrient enrichment, and their interaction on macroalgal proliferation and coral growth. *Coral Reefs*, 28: 555-568.

Tanaka, M. O. and Magalhães, C. A. 2002. Edge effects and succession dynamics in *Brachidontes* mussel beds. *Marine Ecology Progress Series*, 237: 151-158.

Tanaka, M. O. 2005. Recolonization of experimental gaps by the mussels *Brachidontes darwinianus* and *B. solisianus* in a subtropical rocky shore. *Brazilian Archives of Biology and Technology*, 48: 115-119.

Vinueza, L.R.;. Branch, G.M.; Branch, M.L. and Bustamante, R.H. 2006. Top-down herbivory and bottom-up El Ninõ effects on Galápagos rocky-shore communities. *Ecological Monographs, 76: 111-131.*

In: Mussels: Ecology, Life Habits and Control
Editors: Jarek Nowak and Michal Kozlowski
ISBN: 978-1-62618-083-3

Chapter 7

MUSSEL FARMS IN THE GULF OF NAPLES (CAMPANIA, SOUTHERN ITALY): REAL OPPORTUNITY OR A RISK TO CONSUMERS?

Maria Grazia Esposito, Francesca Trinchella and Rosaria Scudiero*
Department of Biology, University of Naples Federico II, Napoli, Italy

ABSTRACT

In recent years, in Italy it has been observed a process of post-industrialization. Many industries closed, and at the same time is occurring a return to more environmentally friendly practices, such as organic farming and responsible fisheries. For the latter the decline of wild aquatic organisms has led to the expansion of aquaculture. Among the molluscs, the edible mussel *Mytilus galloprovincialis* is the most sold in the Southern Italy, and the animals come almost exclusively from aquaculture. The Gulf of Naples is a 15 km coastal area located in the Campania region (Southern Italy), subject to pollution from highly populated urban centers, port activities, and large industrial complexes, including oil refineries. Nevertheless, nowadays many mussel farms have been arising in this area following the closure of many important industries (the steel plant ILVA, for example).

This chapter summarizes the results obtained by monitoring the quality of the mussels *M. galloprovincialis* farmed in different areas of the Gulf of Naples, in terms of heavy metal contamination.

The concentration of toxic heavy metals (Pb, Cd) was determined on the tissues of *M. galloprovincialis* specimens collected at different aquaculture farms. On the same animals was also determined the amount of metallothionein, the most important biomarker of metal pollution. The collected data were compared with those obtained from animals belonging to farms located in pristine areas and from animals relaying in metal free seawater, in order to establish public health risks associated with consuming mussels harvested in the Gulf of Naples.

* Corresponding author: Rosaria Scudiero, Dept. of Biology - University of Naples Federico II, Via Mezzocannone 8, 80134, Napoli, Italy, Tel: +39 0812535217; Fax: +39 0812535035; E-mail: rosaria.scudiero@unina.it.

INTRODUCTION

Extensive cultivation of mussels is a worldwide activity of growing economic importance (FAO, 2004; Prioli, 2008). To meet the increased demand we are assisting to an extension of mussel cultivation to new areas. Although the extent of cultivation was expected in unpolluted areas, in recent years there has been observed an increase of shellfish farming in areas previously occupied by industrial activities. This occurs since the uncontaminated areas are often protected for nature conservation. Moreover, the crisis of industries has led to the closure of many activities that faced the sea (refineries, steel plants, shipyards), releasing many areas available for conversion to more environmentally friendly practices. However, it remains unclear whether the conversion of former industrial areas in other ones used for aquaculture actually represents a real opportunity or if it hides possible risks to consumers. To answer this question, we decided to carry out a screening of the health of mussels cultivated in the Gulf of Naples. This area in fact has experienced this type of conversion; in addition, this area still suffers a high anthropogenic impact. We began our study by analyzing the heavy metals contamination in the mussels' tissues. The results described in this book chapter provide an overview of the quality of *Mytilus galloprovincialis* specimens farmed in the Gulf of Naples, in terms of contamination by heavy metals such as cadmium and lead. The contents of the two metals and the presence of metallothionein, the primary biomarker of metal contamination (Amiard et al., 2006; Trinchella et al., 2006; Scudiero et al., 2008; Cretì et al., 2010), determined both in the digestive and in the reproductive glands, were compared with those measured in animals collected at a reference site and in animals maintained in the laboratory. In particular, for this monitoring it was considered the coast stretching from Capo Miseno to Castel dell'Ovo for about thirty kilometres of coastline. Starting from the North, the mussel farms that have been analyzed are: Capo Miseno (referred to as site 3); Bacoli (referred to as site 4); Nisida (referred to as site 5); Castel dell'Ovo (referred to as site 6). The reference site (referred as site 2) was located in the Gulf of Salerno, that is a more pristine area than the Gulf of Naples (Regione Campania, 2003); finally, the *Mytilus galloprovincialis* specimens kept for 40 days in the laboratory in metal free artificial seawater are referred as site 1. All the animals used for the analyses (90 animals for sites 2-6, 20 animals for site 1) were of commercial size; because differences in growth rate can affect metal concentration in mussels (Ivankovic et al., 2005), analyses were made in mussels of similar size (5-6 cm length). Animals from the site 3 were also collected seasonally from May 2010 to March 2011. Once transferred to the laboratory, mussel shells were open and digestive and reproductive glands were removed from each mussel and divided into two portions (for metals and metallothionein determination), both stored at -20°C until use.

For Cd and Pb determination, tissues were individually digested with concentrated nitric acid (65%, Ultrapure, Fluka) using 1 ml of acid every 50 mg of wet tissue (Cretì et al., 2010). The mixture was heated for 30 min at 90 °C, cooled and centrifuged for 15 min at 12.000x*g*. Cadmium and lead content was determined in three replicates by flame atomic absorption, using a Varian A220 atomic spectrophotometer, according to the manufacturer's conditions. The detection limit of metals in different samples was determined from the standard additions curve for each metal. The detection limit estimated was about 10 ng/g and the recovery, calculated as a percentage of the metals added, ranged from 90% to 93% for both metals. Metallothionein content was determined in each sample using the Micro MT Spec kit (Ikzus

Environment) that is specific to determine the concentration of MT in the mussels' tissues (Viarengo et al., 1997; Trinchella et al., 2012). This protocol is based on the chemical determination of cysteine residues by the 5,5'-Dithio-bis(2-nitrobenzoic acid) (DTNB). Briefly, samples were homogenized in three volumes of homogenizing buffer with 0.1% β-mercaptoethanol and centrifuged at 10,000g for 30 min and at 100,000g for 60 min. The final supernatants were subjected to an ethanol/chloroform extraction followed by an overnight ethanol precipitation at -20°C. The resulting MT-enriched fractions were incubated with DTNB and samples absorbance was read at 412 nm. The interpolation of the obtained value over the glutathione standard curve retrieves the nanomoles of cysteine residues, directly referred to the MTs present in the sample. The conversion of MT molar levels to mass partition (µg/g wet tissue) was calculated on the basis of the estimated molecular mass for the same MT isoform (Trinchella et al., 2012). Statistical analysis of data was performed by using the one-way analysis of variance (ANOVA), the Scheffe post hoc comparison test and the correlation Z test, with the aid of the programs contained in the package StatView (Altura Software, Inc.). Significance level was at 5%.

RESULTS

First, we proceeded to determine cadmium and lead contents in the digestive and reproductive glands of every single animal collected at the different sampling sites (Figure 1). All the samples examined showed measurable concentrations of metals, with the exception of samples from the site 1 for which the content of metals was below the detection limits (Figure 1A). Statistical analyses were carried out on raw data. The results of ANOVA indicated that Cd and Pb levels were significantly higher in the digestive gland with respect to the gonad ($p<0.001$). As regarding differences among specimens from the various aquaculture station examined, our data clearly show that there are no major differences in metals accumulation; pairwise comparisons demonstrate that specimens from the site 2 (Gulf of Salerno) exhibited the highest levels of Cd both in digestive and reproductive glands (average values of 0.17 µg/g and 0.114 µg/g, respectively) (Figure 1B); the lowest level of Cd (average value < 0.03 µg/g tissue) was found in the gonad of the specimens from the site 4 (Nisida station) (Figure 1D). For Pb, a significant higher amount of lead has been found in mussels from the site 5 compared to specimens from the other stations (Figure 1). Data relating to MT (Figure 1) demonstrate a major variability for MT content in the different animals. In general, the MT content was slightly lower in hepatopancreas compared to gonadal tissue ($p=0.028$). However, in mussels maintained in laboratory the MT content is very low and comparable in both the tissues examined (Figure 1A). At the opposite end, a notable amount of MT was found both in digestive and reproductive glands of *M. galloprovincialis* specimens from the Nisida farm station (site 4), followed by animals from the site 6 (Figure 1D, F). Seasonal variability of heavy metals and MT contents in the digestive and reproductive glands of *M. galloprovincialis* has been determined in mussels from the site 3 (Figure 2). Generally, Cd, Pb and MT contents in these mussels were lower than those measured in specimens collected at the reference area (site 2, Gulf of Salerno). In the site 3 (Capo Miseno), heavy metals and MT displayed a statistically significant seasonality ($p<0.05$).

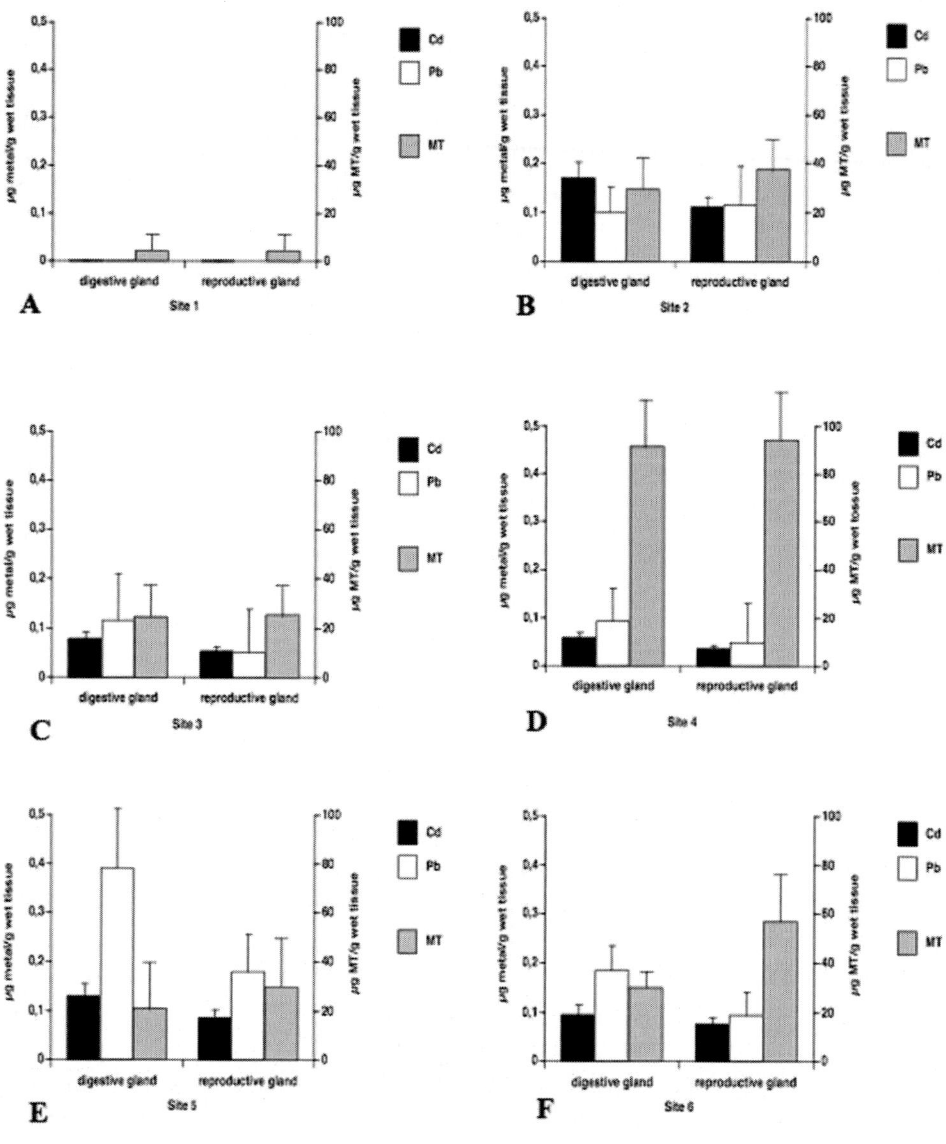

Figure 1. Cadmium (Cd), lead (Pb) and metallothionein (MT) contents, espressed as µg/g of tissue wet weight, in *Mytilus galloprovincialis* digestive gland and reproductive gland of specimens from: laboratory aquarium (A, site 1), reference site (B, site 2) and 4 different mussel farms (C, site 3, Capo Miseno station; D, site 4, Nisida station; E, site 5, Bacoli station; F, site 6, Castel dell'Ovo station). Values are expressed as mean ± S.D. (*n=90*).

The highest MT level in both glands is observed in summer; comparable low values are observed in spring and autumn (beginning of the spawning period). In winter (end of the spawning period) the digestive gland show the lowest MT content, whereas the reproductive gland shows a moderately high amount of MT (Figure 2). In summer, also cadmium highly accumulates in both tissues, followed by a progressive decrease until the spring period (Figure 2). Lead content is extremely low in spring and increases in summer, especially in the digestive gland.

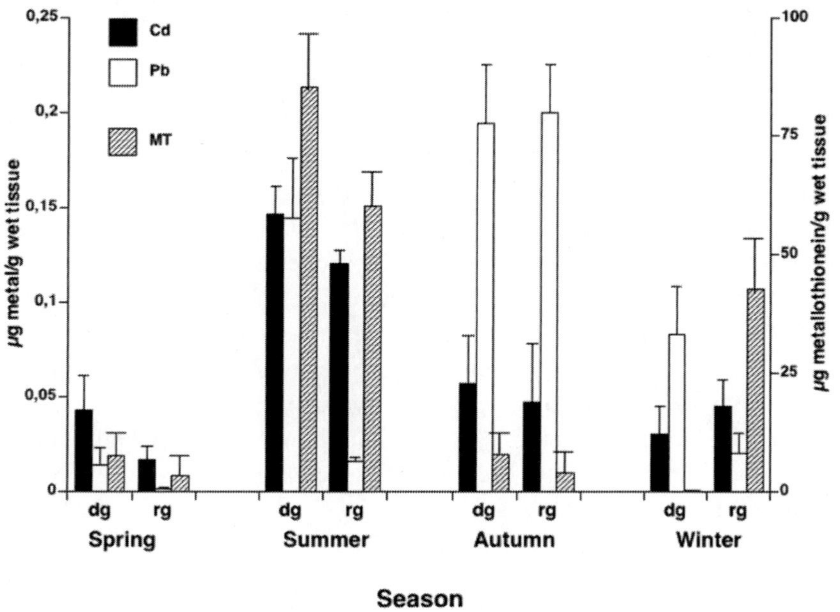

Figure 2. Seasonal variation in cadmium (Cd), lead (Pb) and metallothionein (MT) contents, espressed as µg/g of tissue wet weight, in *Mytilus galloprovincialis* digestive gland (dg) and reproductive gland (rg) of specimens from the site 3, i.e. Capo Miseno station. Values are expressed as mean ± S.D. (*n=20*).

The increase is more pronounced in the autumn period, with the two tissues showing a comparable amount of metal. In winter begins the reduction of the metal amount in both tissues (Figure 2). Correlation analyses demonstrated that the content of both cadmium and lead in mussels' tissues was not strictly correlated with that of MT. Indeed, the high Pb content in digestive gland of mussels from the site 5 is not accompanied by a concomitant accumulation of MT, as well as the high MT amount accumulated in tissues of mussels cultivated at the site 4 is not associated to the presence of Cd and Pb ions (Figure 1).

DISCUSSION

The data collected in this research show that the housing of the mussels in artificial seawater improves the quality of mussels: toxic heavy metals such as cadmium and lead are almost absent and the content of MT, a biomarker of general stress condition, is very low. On the contrary, the comparison among mussels from the various farm stations clearly demonstrates that no great differences were found in the animals in terms of heavy metals and metallothionein contents. However, in all instances the levels of the two metals in both tissues examined were lower than the limits established by the European Food Safety Authority (EFSA 2005), that are 1 µg/g wet tissue for Cd and 1.5 µg/g for Pb. In all the samples examined, also the amounts of MT were quite low, especially if compared with the values measured in tissues of animals grown in polluted areas (Pellerin and Amiard 2009). These results, together with a lack of significant difference in the quality of mussels cultivated in the Gulf of Salerno, demonstrate that mussels from the Gulf of Naples do not present major risks to consumers. Data also demonstrate that metal bioaccumulation varied seasonally, with the highest values during the warmer and rainy periods (summer, autumn), reaching the lowest values in spring. These findings are in accordance with several reports indicating that a series of exogenous and endogenous factors may act on metal bioaccumulation in a complex fashion (Ivankovic et al., 2005). Seasonal variations in metallothionein content and their interpretations have led to much controversy (Domouhtsidou et al. 2004). Metallothionein levels displayed a statistically significant seasonality ($p< 0.05$) with the highest levels in both tissues in summer period, and a high amount in reproductive gland in winter, followed by a progressive decrease during the spring and autumn periods (beginning of the spawning period) in both tissues examined. It is known that several natural factors are able to influence metallothionein biosynthesis. For instance, it has been proposed that seasonal trends of metallothioneins may be related to the reproductive cycle and the growth rate of mussels (Viarengo et al., 1997; Ivankovic et al., 2005). In addition, environmental variables, such as salinity, temperature, and hypoxia are able to increase the cellular concentration of metallothioneins (Riba et al., 2003).

The relatively high MT level found in the reproductive gland could be explained with an intense metabolism, that might bring to the production of a high amount of free radicals that in turn, could induce the MT synthesis (Goshal and Jacob 2001). On the other hand, the low MT content found in the digestive tissue could be due to the fact that heavy metals content did not reached the level that would induce additional MT synthesis. Thus, it is possible to hypothesize that the MT levels measured in all the samples examined were indicative of a low metal exposure. Metal concentration has been positively correlated with metallothioneins in molluscs experimentally exposed to Cd (Serra et al., 1999), and the digestive gland has been often selected as a target tissue for the analysis of MTs as a biomarker of metal exposure in field studies (Lionetto et al., 2001; Andreani et al., 2011). Nevertheless, it is often difficult to establish the same correlation in field specimens due to the variety of environmental and physiological factors influencing MT turnover (Isani et al. 2000; Amiard et al., 2006). The absence of an evident correlation observed in this study between MT content and metal accumulation in the tissues has been already reported for many biological systems (Riggio et al., 2003; Trinchella et al. 2006; Filipovic and Raspor 2007; Cretì et al. 2010). It has been postulated that metal ions are mostly present in tissues not as free ions, but possibly as metal-protein complexes (Moltedo et al., 2000; Trinchella et al. 2006), so the amount of free ions could be too low to bring about a measurable induction of MT expression and synthesis. Therefore, because the concentration of bioavailable heavy metals at highly anthropogenic sites in the Gulf of Naples was comparable to that of the site in the more naturalistic Gulf of Salerno and was lower than the limits established by the European Food Safety Authority, we may conclude that the conversion of industrial sites in areas devoted to aquaculture represents a real opportunity to increase the number of companies and employees in fisheries production.

ACKNOWLEDGMENTS

This study is in the framework of the FARO Project "Aquaculture and Quality (AcQua): characterization of biomarkers for the assessment of coastal waters pollution and the development of mussels farms", financed by the Polo delle Scienze e delle Tecnologie, University Federico II, Naples, Italy.

REFERENCES

Amiard J.C., Amiard-Triquet C., Barka S., Pellerin J., Rainbow, P.S. 2006. Metallothioneins in aquatic invertebrates: their role in metal detoxification and their use as biomarkers. *Aquatic Toxicol.* 76, 160-202.

Andreani G., Carpené E., Capranico G., Isani, G. 2011. Metallothionein cDNA cloning, metallothionein expression and heavy metals in *Scapharca inaequivalvis* along the Northern Adriatic coast of Italy. *Ecotoxicol. Environ. Safety* 74, 366-372.

Cretì P., Trinchella F., Scudiero R.. 2010. Heavy metals bioaccumulation and metallothionein content in tissues of the sea bream *Sparus aurata* from three different fish farming systems. *Environ Monit Assess*, 165, 321-329.

Domouhtsidou, G.P., Dailianis, S., Kaloyianni, M., Dimitriadis, V.K., 2004. Lysosomal membrane stability and metallothionein content in *Mytilus galloprovincialis* (L.), as biomarkers. Combination with trace metal concentrations. *Mar. Pollut. Bull.* 48, 572–586.

EFSA (2005) European Food Sapety Authority. *Annual Report 2004*, pp. 72.

FAO, 2004. FAO Yearbook. Fisheries Statistics: Aquaculture production 2002. Vol. 94/2. Pp. 193.

Filipovic M.V., Raspor B. 2007. Metal exposure assessment in native fish, *Mullus barbatus* L., from the Eastern Adriatic Sea. *Toxicol. Lett.* 168, 292-301.

Ghoshal K., Jacob S.T. 2001. Regulation of metallothionein gene expression. *Prog. Nucl. Acid Res. Mol. Bio.* 66, 357-384.

Isani G., Andreani G., Kindt M., Carpené E. 2000. Metallothioneins (MTs) in marine molluscs. *Cell. Mol. Biol.* 46, 311–330.

Ivankovic, D., Pavicic, J., Erk, M., Filipovic-Marijic, V., Raspor, B., 2005. Evaluation of the *Mytilus galloprovincialis* Lam. Digestive gland metallothionein as a biomarker in a long-term field study: seasonal and spatial variability. *Mar. Pollut. Bull.* 50, 1303–1313.

Lionetto M.G., Giordano M.E., Caricato R., Pascariello M.F., Marinosci L., Schettino, T. 2001. Biomonitoring of heavy metal contamination along the Salento coast (Italy) by metallothionein evaluation in *Mytilus galloprovincialis* and *Mullus barbatus*. *Aquatic Conservation*: *Mar. Freshwater Ecosys.* 11, 305–310.

Moltedo O., Verde C., Capasso A., Parisi E., Remondelli P., Bonatti S., Alvarez-Hernandez X., Glass J., Alvino C.G., Leone, A. 2000. Zinc transport and metallothionein secretion in the intestinal human cell line Caco-2. *J. Biol. Chem.* 275, 31819-31825.

Pellerin J., Amiard J. C. 2009. Comparison of bioaccumulation of metals and induction of metallothioneins in two marine bivalves (*Mytilus edulis* and *Mya arenaria). Comp Biochem. Physiol. C: Pharmacol. Toxicol* 150, 186-195.

Prioli, G. 2008. La molluschicoltura in Italia. In: Lovatelli A., Farías A., Uriarte I. (Eds.), *Estado actual del cultivo y manejo de moluscos bivalvos y su proyección futura: factores que afectan su sustentabilidad en América Latina. Taller Técnico Regional de la FAO.*

20–24 de agosto de 2007. Puerto Montt, Chile. FAO Actas de Pesca y Acuicultura. No. 12. Roma, FAO. pp. 159–176.

Regione Campania. 2003 *Monografia Acque marine Costiere* A6, pp. 40.

Riba, I., Garcia-Luque, E., Blasco, J., DelValls, T.A., 2003. Bioavailability of heavy metals bound to estuarine sediments as a function of pH and salinity values. *Chem. Spec. Bioavail.* 15, 101–114.

Riggio M., Trinchella F., Filosa S., Parisi, E., Scudiero R. 2003. Accumulation of zinc, copper, and metallothionein mRNA in lizard ovary proceeds without a concomitant increase in metallothionein content. *Mol. Repr. Dev.* 66, 374-382.

Scudiero R., Trinchella F., Riggio M., Filosa S. 2008. Trace elements, cadmium and metallothioneins in growing oocytes, eggs and early embryos: a comparative survey on aquatic and terrestrial vertebrates. *Trends Repr. Biol.* 3, 19-29.

Serra R., Isani G., Tramontano G., Carpené E. 1999. Seasonal dependence of cadmium accumulation and Cd-binding proteins in *Mytilus galloprovincialis* exposed to cadmium. *Comp Biochem. Physiol. C: Pharmacol. Toxicol.* 123, 165–174.

Trinchella F., Esposito M.G., Simoniello P., Scudiero R. 2012. Cadmium, lead and metallothionein contents in cultivated mussels (*Mytilus galloprovincialis*) from the Gulf of Naples (Southern Italy). *Aquaculture Res.* Epub 2012 DOI: 10.1111/j.1365-2109.2012.03111.x.

Trinchella F., Riggio M., Filosa S., Volpe M. G., Parisi E., Scudiero R. 2006. Cadmium distribution and metallothionein expression in lizard tissues following acute and chronic cadmium intoxication. *Comp. Biochem. Physiol. C Toxicol. Pharmacol.* 144, 272-278.

Viarengo A., Ponzano E., Dondero F., Fabbri R. 1997. A simple spectrophometric method for metallothionein evaluation in marine organisms: an application to Mediterranean and Antarctic molluscs. *Mar. Environ. Res.* 44, 69-84.

INDEX

#

20th century, 9

A

access, 87
acetic acid, 52
acetylcholinesterase, 86
acid, 52, 55, 61, 65, 68, 69, 70, 71, 72, 74, 75, 150
actual growth rate, 118
acylation, 63
adaptability, 88
adaptation(s), 26, 88, 97
additives, 83
adductor, 5, 47, 54, 57, 58
adhesion, 52
adsorption, 19
adulthood, 53
adults, 10, 11, 26, 88, 89, 142, 144
adverse conditions, 11
adverse effects, 39, 64
Africa, 9, 52, 56
age, 74, 88
agencies, 14, 40
agriculture, 3, 6, 22, 25, 39
Agrobacterium, 24
air temperature, x, 137, 139, 142
Alaska, 33
Alexandrium catenella, viii, 37, 43, 44, 45, 47, 49, 62, 66, 67
algae, 7, 10, 13, 15, 18, 19, 46, 49, 53, 55, 56, 62, 138, 139, 141, 143, 144, 146, 147
algal sub-environments, x, 137, 141, 142, 146
alien species, 27, 33
allometry, 111
alters, 73
aluminium, 81
amino, 6
amino acid(s), 6
ammonia, 4, 17
anatomy, 3
ANOVA, 144, 151
antibody, 94, 126, 131
antioxidant, 85, 94
antisense, 126, 129
aquaculture, viii, xi, 2, 3, 12, 13, 25, 37, 38, 39, 40, 53, 61, 62, 63, 67, 149, 150, 151, 154
aquaria, 104
aquarium, 152
aquatic environments, vii, 1, 20, 27, 79, 82, 83, 86, 134
aquatic habitats, 19, 30
aquatic life, 3, 27
aquatic systems, 20, 21, 25
Argentina, 8, 15, 38, 44, 52
aromatic hydrocarbons, 4, 33, 93, 94
Asia, 4, 8, 12, 15, 22, 23, 95, 98
Asian countries, 22
assessment, ix, 35, 62, 65, 66, 72, 77, 78, 79, 87, 89, 90, 91, 92, 97, 98, 102, 154, 155
ataxia, 43
atmosphere, 21
ATP, 53
at-risk populations, 78
attachment, 11, 13, 87, 102
Austria, 9
awareness, 78

B

bacteria, viii, 2, 10, 15, 16, 17, 19, 21, 24, 38, 46, 63, 68, 83, 94, 98
bacterial infection, 24

bacterial pathogens, 30
bacterial strains, 24
ban, 82, 94
banks, 44, 62
barriers, 25, 79
base, 46, 67, 116
behaviors, 62
Belgium, 52, 94, 135
beneficial effect, 4
benefits, 7, 25
benthic invertebrates, 19
bioaccumulation, 20, 27, 32, 33, 35, 78, 99, 124, 132, 154, 155
bioassay, 43, 63, 94
bioavailability, 28, 78
biochemical processes, 21
bioconversion, 57, 58, 72, 73
biodiversity, 27
bioindicators, viii, 2, 20, 21, 22, 29, 30, 32, 95, 134
bio-indicators, 79, 80
biological activities, 13
biological processes, 20
biological responses, 78
biological systems, 82, 154
biomarkers, 29, 74, 78, 81, 84, 85, 86, 87, 89, 90, 91, 93, 94, 95, 97, 99, 124, 134, 154, 155
biomass, x, 4, 7, 18, 19, 26, 30, 39, 40, 121, 137, 138, 139, 142, 143, 145, 146
biomonitoring, ix, 21, 23, 77, 78, 79, 80, 81, 84, 85, 87, 88, 89, 90, 93, 94, 96, 97, 99, 124, 125
biosynthesis, 154
biotic, 25, 31, 79
bisphenol, 83, 133
blindness, 79, 82
blood, 53, 70, 79
bloodstream, x, 123, 124
body fluid, 8, 66, 79
body weight, 24, 88
bottom-up, 138, 147, 148
Brazil, v, vii, 22, 38, 44, 97, 99, 137, 138, 139, 146, 147, 148
breakdown, 69
breast cancer, 99
breeding, 88, 89
Britain, 16
buttons, vii, 1, 9
by-products, 83, 85

C

cadmium, 81, 94, 150, 151, 152, 153, 156
calcium, 10, 11, 14, 15, 16, 26
calcium carbonate, 10
Campania region, xi, 149
canals, vii, 1, 3, 7, 9, 10
cancer, 17, 23, 31, 52, 69, 82
carbon, 22
carcinoma, 70
cascades, 52
case study, 33
Caspian Sea, 7, 33
catchments, 5
cDNA, 125, 129, 134, 155
cell line(s), 56, 73, 155
cell metabolism, 24
cell organelles, 85
Central Europe, 28
central nervous system, 134
chemical(s), viii, ix, 3, 7, 8, 17, 19, 20, 21, 22, 23, 27, 28, 38, 40, 46, 47, 49, 54, 57, 58, 61, 62, 64, 77, 78, 79, 80, 81, 82, 84, 86, 87, 88, 89, 94, 95, 96, 98, 125, 151
chemical properties, 40
chemical stability, 58
Chicago, 28, 29
Chile, 37, 38, 44, 52, 54, 55, 56, 62, 64, 66, 67, 68, 70, 72, 73, 75, 156
China, 9, 12, 13, 29, 72
chlorine, 25
chloroform, 151
chlorophyll, 105, 106, 114, 116, 118
chromatid, 84, 92
chromatography, 66
chromosome, 84
cilia, 13, 15
circulation, 5, 39, 44
classes, x, 5, 40, 74, 117, 118, 119, 137, 139, 141, 142, 143
classification, 24, 78
cleaning, 7
climate(s), 38, 138
climate change, 38
cloning, 134, 155
closure, x, xi, 87, 137, 141, 144, 149, 150
CO_2, 52
coatings, 26
colon, 56, 70
colon carcinogenesis, 70
colonization, vii, x, 137, 138, 139, 141, 142, 143, 144, 145, 146, 147
color, 16
colorectal cancer, 69
combustion, 29, 31, 83
commercial, 3, 6, 13, 39, 68, 150

community(s), ix, 7, 8, 9, 18, 19, 20, 23, 26, 27, 28, 39, 40, 77, 78, 79, 81, 88, 89, 97, 138, 139, 141, 146, 148
compensation, 88
competition, ix, 28, 39, 79, 101, 103, 117, 138, 144, 145
competitors, 138
complex interactions, 84
complexity, 19, 26, 138
compliance, 147
composition, 8, 14, 27, 40, 43, 44, 46, 66, 71, 74, 120
compounds, viii, 2, 17, 20, 22, 32, 37, 38, 53, 70, 72, 82, 83, 84, 85, 88, 93, 96, 134
conceptual model, 64
conductance, 42
configuration, 61
Congress, 67
conjugation, 85
connective tissue, 124, 129, 130
conservation, 10, 27, 94
construction, 16
consumers, 25, 52, 68, 150, 154
consumption, 4, 23, 24, 30, 49, 52, 65, 69
contaminant, 32, 78, 80, 84, 85, 88, 97, 121
contaminated sites, 86
contaminated water, 16, 84
contamination, xi, 17, 20, 21, 24, 32, 34, 52, 69, 81, 82, 83, 84, 85, 89, 90, 91, 94, 95, 98, 99, 149, 150, 155
Continental, 91
control measures, 23
controversial, 12
cooking, 65
cooling, 82, 87
copper, 94, 95, 135, 156
copyright, 133
coral reefs, 146, 147
correlation(s), ix, x, 23, 67, 101, 107, 115, 137, 142, 143, 145, 146, 151, 154
correlation coefficient, 116, 142
corrosion, 7
cortex, 128
cost, 26, 27, 34, 78, 87
covering, ix, 12, 25, 101, 118
crabs, 39, 135
Croatia, 55
crop(s), 39, 40
crude oil, 86
cultivation, 10, 150
culture, 30, 93, 97, 120
currency, 6
cycles, 14, 22, 125, 134
cycling, 17, 25, 27, 65
cyclooxygenase, 92
cyst, 43, 67
cysteine, 86, 151
cysteine-rich protein, 86
cytochrome, 85, 96
cytoplasm, 132
cytoskeleton, 56
Czech Republic, 9

D

damages, 25, 27
danger, 6
data analysis, 102, 141
data collection, 107
data set, 116
DDT, 2, 22, 23, 28, 33, 78, 82, 85, 90, 91, 98
decay, 19
decision makers, 78
defence, 86, 87
defense mechanisms, 57
deficiency(s), 78, 89
degradation, 17, 65, 78, 88
Delta, 98
Denmark, ix, 52, 101, 103, 104, 115, 120
density values, 142
dephosphorylation, 52
deposition, 20, 93, 144
depth, 10, 15, 29, 58, 61, 75
derivatives, 22, 44, 46, 52, 53, 54, 57, 65, 70
desiccation, 80, 138, 142, 145
destruction, 16, 53, 82
detectable, 127
detection, 26, 32, 40, 49, 58, 63, 66, 73, 150, 151
detoxification, 52, 56, 61, 67, 74, 85, 86, 155
developed countries, 28, 91
developing countries, 82, 98
diarrhea, 24, 54, 56
diatoms, 17, 18, 39
diet, viii, 12, 37, 38, 47, 49, 58, 59
diffusion, 54
digestion, 5, 13, 49, 53
dioxin, 24, 30
directionality, 67
directives, 24
discharges, 94
discrimination, 87
diseases, 78
displacement, 26
disposition, 78
dissolved oxygen, 14, 15, 25

distribution, ix, 7, 11, 14, 15, 25, 26, 29, 30, 32, 44, 47, 48, 57, 59, 60, 62, 65, 67, 68, 74, 75, 77, 80, 87, 88, 89, 98, 121, 124, 132, 139, 141, 142, 144, 145, 148, 156
diversity, 3, 9, 13, 16, 18, 19, 66, 79, 81, 88, 89, 138
dizziness, 43
DNA, 34, 80, 84, 85, 90, 92, 99, 126
DNA damage, 80, 84, 85, 90, 92, 99
DOI, 156
dominance, 58, 138
drainage, 16, 88
drinking water, 3
drought, 20
drugs, 17, 20, 53
drying, 104
dysarthria, 43
dysphagia, 43
dyspnea, 54

E

early warning, 79, 84
Eastern Europe, 7
E-cadherin, 56, 73
ecological systems, 95
ecology, vii, 18, 30, 32, 35, 73, 93, 97, 103, 119, 120, 121
economic damage(s), 25, 27
economic losses, 25
economic problem, 8
economics, 31
ecosystem, vii, viii, 1, 2, 4, 18, 26, 27, 30, 34, 35, 103, 147
ecotoxicological, 89
ecotoxicology, x, 124
editors, 133
effluent(s), 3, 18, 20, 82, 86, 98, 134, 135
egg, x, 86, 102, 123, 134
electron, x, 124, 125, 126, 127, 128
electron microscopy, 125, 126
ELISA, 71
empirical studies, 26
employees, 154
encoding, 134
encyst, 11
endangered, 39
endocrine, x, 78, 82, 83, 86, 89, 92, 94, 96, 123, 124, 132, 133, 134, 135
endocrine system, 78, 83
energetic parameters, 103
energy, 20, 25, 27, 31, 43, 83, 102, 103, 120, 121
engineering, 4, 18
England, 20, 29, 32, 119
environment(s), vii, ix, x, 1, 3, 4, 7, 11, 14, 18, 20, 21, 22, 23, 24, 25, 26, 27, 29, 38, 40, 53, 63, 77, 78, 79, 81, 82, 83, 85, 86, 87, 88, 89, 95, 134, 137, 138, 139, 140, 141, 142, 143, 144, 145, 146
environmental change, viii, 2, 87, 89
environmental conditions, 4, 8, 10, 12, 14, 15, 16, 43, 80, 84, 119
environmental contamination, 21, 22, 99
environmental factors, x, 25, 88, 102, 137, 139
environmental impact, 30, 41
environmental management, 98
environmental movement, 78
Environmental Protection Agency, 89
environmental quality, 30, 80, 87
environmental stress, 80, 89, 99, 145
environmental variables, 15, 154
enzyme(s), 21, 40, 52, 53, 54, 63, 80, 85
EPA, 28, 31, 89
epithelial cells, 73
equipment, 125
ester, 53, 54, 55, 68, 72
estrogen, 86, 92, 124, 132, 133, 135
estuarine environments, 83
ethanol, 125, 126, 151
EU, 49, 56, 65
eukaryotic, 52
eukaryotic cell, 52
Eurasia, 15
Europe, vii, 1, 3, 4, 7, 9, 12, 14, 22, 31, 40, 51, 55, 80
European Commission, 2, 65, 69
European Union, 70
Everglades, 88
evidence, x, 7, 15, 16, 29, 56, 58, 68, 73, 75, 78, 86, 124, 129, 130, 135
evolution, 64, 93
excretion, 18, 19, 53, 63, 70
exploitation, 17, 62, 102
exports, 62
exposure, ix, 3, 20, 23, 24, 38, 46, 52, 53, 58, 64, 66, 69, 77, 78, 79, 80, 81, 83, 84, 85, 86, 87, 88, 89, 90, 92, 96, 99, 132, 134, 135, 154, 155
extinction, 9, 16, 39
extraction, 17, 38, 39, 64, 86, 125, 151
extracts, 56, 62, 65

F

families, vii, 1, 8, 17, 85, 90
farmers, 39
farms, vii, xi, 149, 150, 152, 155
fat, 51
fatty acids, 52, 63

fauna, 8, 25, 39, 141
feces, 19
fertilization, 14, 16
fertilizers, 16, 77
fever, 24
filters, 27
filtration, ix, 10, 19, 25, 29, 35, 46, 54, 61, 62, 64, 101, 102, 119, 121
financial, 27, 146
financial support, 27, 146
Finland, 121
fire retardants, 23
fish, x, 7, 9, 10, 11, 16, 17, 18, 21, 23, 24, 27, 35, 45, 67, 79, 85, 86, 90, 94, 123, 124, 132, 134, 155
Fish and Wildlife Service, 3, 30, 34
fisheries, xi, 149, 154
fishing, 6, 9, 16, 39
fitness, 88
fixation, 126
flame, 82, 150
fluctuations, 80
fluid, 24, 87
fluorescence, 66, 73
follicle(s), x, 124, 125, 127, 128, 129, 130, 131, 132, 133, 134, 135
food, vii, viii, ix, 1, 2, 5, 6, 7, 9, 10, 12, 13, 14, 16, 17, 18, 19, 20, 21, 22, 23, 24, 25, 26, 27, 28, 32, 35, 38, 39, 40, 47, 49, 52, 58, 60, 61, 68, 69, 70, 74, 78, 79, 80, 81, 82, 87, 89, 97, 101, 102, 103, 120
food chain, viii, 2, 7, 12, 21, 22, 23, 25, 26, 49, 52, 69, 74, 81, 82, 89
food products, 23
food web, 17, 20, 32, 35, 68, 79, 89
force, 23
formation, 43, 44, 63, 84, 85, 97
formula, 102
fouling, 33
fragments, 53, 84, 129
France, 12, 28, 29, 32, 52, 56, 69, 92, 98
free radicals, 154
freezing, 15, 52
freshwater, vii, viii, 1, 2, 3, 4, 5, 6, 7, 8, 9, 10, 11, 14, 16, 17, 19, 21, 22, 25, 26, 27, 28, 29, 32, 33, 34, 35, 82, 87, 88, 91, 93, 94, 95, 96, 97, 134
freshwater habitats, vii, 1, 4, 25
freshwater species, viii, 2, 16, 21
friction, 7

G

gametogenesis, 86, 88, 92, 134
gastrointestinal tract, 19
gel, 125
gene expression, 52, 69, 135, 155
genes, 92, 99
genetic diversity, 88, 89
genetics, 30, 93, 97
genus, 11, 21, 24, 26, 31, 43, 64, 143
geography, 62
Germany, 125
germination, 44, 67
gifted, 126
gill, 53, 54, 88, 97
gland, x, 61, 74, 85, 124, 125, 126, 129, 130, 131, 132, 133, 151, 152, 153, 154, 155
glutathione, 85, 151
glycogen, 53
gonads, 47, 57, 58, 92, 126, 129, 132
grain size, 18
granules, 126
grazing, 67, 120
green alga, 18, 139, 144
growth, vii, ix, 7, 11, 12, 13, 15, 18, 20, 39, 43, 44, 53, 61, 66, 80, 87, 101, 102, 103, 104, 106, 107, 108, 109, 110, 111, 112, 113, 114, 115, 116, 117, 118, 119, 120, 121, 124, 129, 135, 138, 139, 144, 148, 150, 154
growth rate, ix, 11, 12, 44, 66, 101, 102, 103, 106, 107, 108, 109, 110, 111, 112, 113, 114, 115, 116, 117, 118, 150, 154
guidelines, 49

H

habitat(s), vii, viii, ix, 1, 4, 7, 8, 9, 11, 12, 13, 16, 17, 18, 19, 25, 27, 37, 38, 46, 47, 49, 57, 58, 62, 63, 87, 88, 146, 147
HAP, 77
harbors, 7, 12
harmful effects, 24, 40
harmonization, 90
harvesting, 39
Hawaii, 13
hazardous substances, 3
head lice, 82
headache, 43
health, x, 3, 4, 8, 18, 23, 27, 30, 35, 74, 79, 88, 90, 123, 150
health effects, 23, 90
heart rate, 88
heat shock protein, 134
heavy metals, viii, xi, 2, 3, 4, 17, 19, 21, 22, 29, 30, 31, 34, 81, 85, 86, 87, 91, 92, 95, 99, 149, 150, 151, 153, 154, 155, 156
heterogeneity, 19, 141

hexane, 65
high heavy metals, viii, 2
history, 11, 28, 106, 115, 141
homeostasis, 86
Hong Kong, 8, 22, 31, 95
hormones, x, 123, 124
host, 10, 11, 24, 35, 86
housing, 153
human, vii, viii, 1, 2, 3, 4, 7, 9, 13, 17, 20, 21, 23, 24, 25, 30, 34, 37, 39, 45, 49, 53, 54, 55, 63, 64, 65, 66, 68, 70, 71, 73, 78, 79, 83, 88, 89, 94, 97, 155
human body, 63
human development, viii, 21, 37, 39
human health, 34, 55, 63, 64, 66, 68, 73, 78, 83, 89, 94
hybrid, 12, 80, 91
hybridization, 126, 129, 130
hydrocarbons, 92
hydrolysis, 46, 85, 86
hydrophobicity, 58, 63
hydroxyl, 46, 52
hypothesis, 89, 107, 133, 139
hypoxia, 154

I

ideal, ix, 77, 80, 89, 133
identification, 12, 47, 57, 64, 66, 70, 74, 78
identity, 35
images, 133
immigration, 88
immune system, 23, 86
immunity, 99
immunohistochemistry, x, 124, 125, 132, 133
immunomodulation, 96
in situ hybridization, x, 124, 125, 130, 132
in vitro, 54, 55, 56, 85, 126
in vivo, 53, 55, 85
incidence, 86
India, 22, 23, 30, 33, 34, 98
individuals, x, 9, 10, 14, 15, 16, 19, 63, 80, 88, 104, 107, 108, 109, 110, 137, 141, 142, 143, 144, 146
induction, 85, 90, 134, 135, 154, 155
industrialization, xi, 149
industry(s), xi, 6, 7, 8, 9, 21, 40, 51, 149, 150
infancy, 26
infection, 24
infrastructure, 25
ingest, 10
ingestion, 12, 23, 51, 55, 56, 103
ingredients, 70
inhibition, 54, 86, 97
inhibitor, 70
insecticide, 79
insects, 18
insertion, 40
integration, 87, 106, 115
integrity, 3, 13
interface, 10
international standards, 61
interrelations, 93
intersex, 86
intervention, 78
intoxication, 24, 54, 56, 57, 65, 70, 156
invertebrates, x, 7, 10, 17, 18, 26, 30, 63, 79, 86, 121, 134, 138, 147, 155
investments, 27
ionizable groups, 54
ionization, 70
ions, 81, 153, 154
Ireland, 13, 52, 55, 56, 71, 74
iron, 96
irradiation, 52
isolation, 73
isomers, viii, 38, 41, 45, 46, 47, 52, 55, 57, 58, 62, 63
isozyme, 85
issues, 18, 40, 126
Italy, v, xi, 24, 28, 32, 52, 54, 91, 93, 123, 134, 149, 155, 156

J

Jamaica, 147
Japan, 4, 6, 8, 12, 31, 34, 52, 55, 67, 70, 126
juveniles, 10, 11, 17, 26, 28

K

kidney, 56
kill, 6, 19, 24
kinetics, 67, 68, 81
Korea, 12, 13

L

labeling, 126
laboratory studies, 102
lakes, vii, 1, 3, 7, 9, 10, 17, 20
larva, 102
larvae, 7, 11, 13, 15, 26, 61, 88, 89, 107, 108, 119, 120, 121, 138
larval stages, 5, 61, 80
Latin America, 23
laws, 103

LC-MS, 49, 54, 56, 68, 71
LC-MS/MS, 49, 56, 71
lead, 14, 16, 18, 33, 43, 52, 63, 88, 90, 116, 126, 150, 151, 152, 153, 156
lead content, 150, 151
legislation, 40
lesions, 80
life cycle, 3, 10, 11, 78, 79
ligament, 5
light, 15, 18, 43, 82
limestone, 11
lipids, 58
lipophilic toxins, vii, viii, ix, 38, 40, 49, 57, 58, 59, 61, 62, 71
liquid chromatography, 54, 66, 70, 71, 72, 73
liquids, 82
lithium, 81
liver, 85, 124
livestock, 6
longevity, 11, 16
low temperatures, 57
lumen, 127, 132
luminescence, 62
lymphocytes, 71
lysozyme, 86

M

macroalgae, 138, 141, 147
magnitude, 20, 78, 116
majority, 3, 8, 22, 53, 56
malaria, 82
Malaysia, 22, 31
mammal(s), 39, 45, 56
man, viii, 16, 20, 37, 39, 92
management, viii, 2, 19, 26, 27, 32, 66, 78, 83, 94, 96
manganese, 96
manipulation, 26
mantle, 5, 47, 53, 54, 58, 87, 88
manufacturing, 13, 64
marine environment, 21, 29, 89, 96, 98, 99
marine species, 13, 39
mass, ix, 5, 54, 66, 70, 72, 73, 101, 151
mass spectrometry, 54, 66, 70, 72, 73
materials, 25, 83, 128, 147
matrix, 40, 57, 82
matter, 19, 20, 49, 119
measurement(s), 32, 65, 71, 78, 80, 81, 86, 87, 92, 102, 109, 111, 141
meat, 6, 14, 23, 24, 57, 62, 115
media, 78
median, 23
medical, 6
medicine, 12
Mediterranean, viii, x, 2, 4, 7, 11, 12, 13, 23, 30, 67, 69, 82, 92, 98, 123, 124, 125, 144, 148, 156
membranes, 53
mercury, 88
Metabolic, 65, 121
metabolic pathways, 40, 45
metabolise xenobiotics, ix, 77, 89
metabolism, 46, 49, 55, 57, 71, 72, 78, 85, 87, 102, 154
metabolites, 43, 72, 74, 79, 82, 83, 85
metabolized, 55
metabolizing, 20, 63, 80
metal ion(s), 154
metals, viii, ix, 2, 4, 19, 21, 22, 28, 29, 31, 34, 77, 81, 85, 86, 87, 89, 93, 95, 96, 97, 98, 150, 151, 153, 154, 155
metamorphosis, 102, 119
Mexico, 8, 13, 33, 138
mice, 54, 56, 57, 71
micrometer, 141
micronucleus, 99
micronutrients, viii, 37, 38
microorganisms, 6, 17, 32
microscope, 104, 126, 127
microscopy, x, 124
migration, 61, 141, 144, 146
Mississippi River, 9
mitosis, 84
models, ix, 69, 77, 103, 120, 138
modifications, 16, 40, 46, 54, 64
molecular mass, 151
molecular weight, 40
molecules, x, 53, 85, 123
mollusks, vii, 1, 5, 8, 17, 19, 23, 25, 67, 139
Montenegro, 30, 31
Morocco, 13, 52, 56
morphology, 3, 5, 120, 121
mortality, 7, 17, 53, 69, 88, 94, 138, 144
mortality rate, 144
Moscow, 35
mRNA, 86, 90, 92, 129, 130, 133, 156
mtDNA, 90
mucus, 13
multivariate data analysis, 94
municipal solid waste, 35
muscles, 5
muskrat, 17
Mussel Watch Program, 23, 80
mutagenesis, 67
mutation(s), 46, 65, 67

N

naming, 5
native species, 9, 25, 27
natural resources, 39, 40
nature conservation, 150
nausea, 24, 43
necrosis, 80
negative consequences, 87
negative effects, 117, 138
nerve, 42, 67
nervous system, 23
Netherlands, 34
neurons, 65
New Zealand, 11, 31, 54, 55, 72
nickel, 32
nitrogen, 16, 43, 44
NMR, 54
NOAA, 2, 23, 29
non–mobile shellfish, vii, 1
North America, 4, 6, 7, 9, 11, 12, 14, 15, 16, 19, 22, 28, 31, 32, 33, 34, 35
Norway, 25, 33, 52, 54, 55, 56, 71, 72, 94, 134, 135
NRCS, 2, 6, 8, 9, 10, 17, 19, 32
nucleotide sequence, 126, 129
nuisance, 8, 26, 31
nutrient(s), 7, 14, 17, 18, 19, 20, 21, 27, 28, 39, 40, 43, 44, 62, 65, 68, 148
nutrition, 117

O

obstruction, 7
occlusion, 7
oceans, 3, 20
OH, 55
oil, xi, 34, 86, 88, 93, 95, 133, 149
oil spill, 34, 88, 93, 95
oocyte, x, 124, 125, 127, 128, 129, 132, 133, 135
oogenesis, 134
operations, 3
opportunities, 98
organ(s), 5, 13, 57, 63, 74, 85, 86
organelles, 53, 126
organic compounds, viii, 2, 4, 17, 29, 93
organic matter, vii, 1, 7, 12, 13, 18, 19, 20, 83
organism, ix, x, 8, 19, 24, 53, 58, 67, 77, 78, 80, 81, 84, 85, 86, 87, 89, 90, 123, 138
organotin compounds, 98
oscillation, 44
osmium, 126
ovaries, 125, 126
overlap, 12
ox, 10, 11, 81, 87
oxidation, 43, 54, 56, 82, 85
oxidative stress, 85, 86, 94
oxygen, 5, 10, 12, 15, 16, 25, 61, 80, 102, 120, 121
oxygen consumption, 120, 121
oyster(s), 4, 5, 6, 22, 23, 28, 29, 43, 67, 69, 95, 96, 119, 134, 138
ozonation, 52, 53
ozone, 53

P

Pacific, v, 11, 22, 37, 38, 75, 95, 98, 119, 134
pain, 56
paints, 82, 93
paradigm shift, 66
parallel, 141
paralysis, 41, 56
parasites, 30, 34
paresthesias, 43
Parliament, 70
partition, 151
pathogens, 4, 20, 24, 52
pathways, viii, 38, 40, 46, 47, 49, 54, 63, 87
PCBs, 20, 23, 24, 29, 33, 82, 85, 88, 92
PCDD/Fs, 2, 24
PCR, x, 90, 124, 125, 129, 132, 133
Pearl River Delta, 29
peptide, 18
peripheral nervous system, 86
permission, 133
Persistent Organic Pollutants, 2
personal communication, 116
pesticide, 82
pests, 52
petroleum, 22, 81, 92
pH, 8, 14, 15, 16, 26, 44, 54, 126, 156
phagocytosis, 86, 91
pharmaceutical(s), 77, 89, 94
phase shifts, 138
Philippines, 23, 98
phosphate, 86, 124, 126, 134
photosynthesis, 43, 66
phycotoxins, viii, 37, 39, 40, 62, 66, 75
phylum, 5, 8
physical characteristics, 16
physical structure, 18
Physiological, 74, 79, 80, 87, 88, 93, 120, 121
physiological factors, 154
physiology, 3, 30, 32, 84, 86, 90, 93, 97, 120, 121, 148

phytoplankton, viii, 10, 13, 15, 16, 17, 18, 28, 35, 37, 38, 39, 40, 48, 49, 60, 64, 81
pilot study, 32
plankton, 13, 69, 72, 73
plant growth, 18
plants, 8, 18, 19, 25, 26, 83, 150
plasma membrane, 53
plasticity, 14, 144
plastics, 6, 82
platelets, x, 123, 124, 126, 127, 128, 132
plutonium, 83
poison, 55
polar, 85
polarity, 85
pollutants, viii, 2, 3, 4, 6, 11, 17, 19, 20, 21, 22, 23, 27, 33, 35, 78, 79, 80, 82, 84, 85, 86, 88, 89, 90, 94, 95
pollution, vii, viii, ix, x, xi, 1, 2, 3, 6, 7, 9, 13, 16, 17, 20, 21, 22, 27, 28, 31, 32, 33, 34, 35, 77, 78, 79, 80, 81, 83, 84, 85, 86, 87, 88, 89, 90, 91, 92, 93, 95, 96, 97, 98, 99, 103, 123, 124, 125, 132, 133, 149, 154
polybrominated diphenyl ethers, 23
polycarbonate, 83
polychlorinated biphenyl, 29, 31
polycyclic aromatic hydrocarbon, 22, 33, 34, 68, 83, 85
polyether, 51, 54, 55, 68, 73
polyhalogenated biphenyls, 23
ponds, 3, 9
pools, 5
POPs, 20, 22, 23, 24, 32, 35
population, 7, 14, 15, 16, 18, 25, 27, 30, 39, 44, 61, 62, 67, 80, 81, 88, 89, 93, 103, 116, 119, 120, 121, 138, 141, 144, 146, 147
population density, 81
population growth, 25, 44
population size, 93
population structure, 121, 138, 146
Portugal, 52, 55, 56, 66, 72, 144
positive correlation, x, 137, 142
positive relationship, x, 137
power plants, 7, 8, 25, 26
precipitation, 151
predation, 25, 79, 97, 145
predators, viii, 2, 5, 11, 54, 61
preparation, 104
prevention, 27, 40, 78
Prince William Sound, 33
probability, 64, 78
probe, 129
project, 119
proliferation, 39, 85, 91, 148
promoter, 69, 82
propylene, 126
protection, 20, 83, 86
protein synthesis, 132
proteins, x, 52, 80, 92, 96, 123, 124, 133, 134, 156
proteomics, 90
PST, 46
public concern, 78
public health, xi, 51, 57, 72, 149
public interest, 17
purification, ix, 17, 38, 47, 58, 60, 61, 64

Q

quality control, 129
quality improvement, 3, 19
quantification, 64, 86

R

radiation, 52, 70, 83, 87, 90
radioactive waste, 83
radium, 83
radula, 5
rainfall, x, 137, 139, 140, 142, 145
RE, 121
reactions, viii, 38, 46, 85, 87
reactive oxygen, 85
receptors, 52, 65
recovery, x, 137, 146, 150
recreation, 17, 25, 40
redistribution, 99
regression, 108, 111, 115, 116, 144
regression analysis, 144
regression line, 108, 111, 115, 116
regulations, 40, 43, 78, 83
regulatory agencies, 78
relatives, viii, 2, 5, 21
relevance, 97, 124
reliability, 21
repellent, 82
reproduction, 5, 14, 26, 80, 90, 96, 119, 134
requirements, viii, 14, 37, 39, 43
researchers, 21, 25
residues, 32, 86, 91, 151
resistance, 46, 53, 65, 67, 70
resources, viii, 27, 37, 38, 39, 40, 61, 62, 64, 96
respiration, ix, 5, 101, 102, 103, 119, 120
respiratory disorders, 56
respiratory failure, 42
response, 43, 53, 67, 78, 87, 90, 91, 92, 94, 103
restrictions, 23, 82

retail, 24
RH, 119
ribosomal RNA, 129
risk(s), ix, xi, 11, 25, 30, 34, 35, 40, 52, 55, 64, 65, 69, 73, 77, 78, 79, 86, 87, 89, 90, 91, 97, 99, 149, 150, 154
risk assessment, 35, 78, 79, 91, 99
risk management, 64, 78
RNA, 125, 126
room temperature, 126
root, 141
roughness, 148
routes, 27
runoff, 3, 7, 16, 20
Russia, 7, 9, 14, 22, 54

S

safety, 66, 74
saline water, 98
salinity, 8, 13, 14, 15, 25, 43, 44, 60, 66, 80, 102, 104, 105, 114, 116, 118, 121, 138, 145, 148, 154, 156
salinity levels, 8
salmon, 10, 11
Salmonella, 24
saltwater, vii, 1, 4, 8
SAS, 141
saturation, 103, 121
scaling, 120
scaling law, 120
Scandinavia, 9
scatter, 116
science, 5, 27
scope, 103, 120, 148
seafood, viii, 2, 34, 43, 49, 63, 94
seasonality, 139, 151, 154
Second World, 6
secrete, 5
secretion, 155
sediment(s), 3, 4, 6, 10, 11, 12, 18, 19, 21, 23, 27, 30, 38, 80, 81, 88, 91, 93, 94, 98, 144, 156
sedimentation, 16
seed, 74
segregation, 13
sensation, 43
sensitivity, 17, 21, 46, 94
Serbia, 1
serine, 52
serotonin, 86, 92
serum, 66
sewage, 81, 86, 95
sex, 91, 124
shape, 5, 43, 126, 128, 132
shellfish, vii, viii, ix, 1, 5, 6, 38, 40, 41, 43, 46, 49, 51, 52, 54, 55, 56, 57, 58, 61, 62, 63, 64, 65, 66, 67, 68, 69, 70, 71, 72, 73, 74, 80, 150
shock, 52
shorelines, 33
shores, vii, 1, 3, 12, 31
shortness of breath, 43
showing, 46, 47, 63, 102, 104, 138, 153
signalling, 90
signs, 79
siphon, 5, 13, 15
sister chromatid exchange, 84, 96
skin, 69
sleeping sickness, 82
small intestine, 57
social development, 40
sodium, 41, 46, 67
software, 141
solution, 125
somatic cell, 129, 130
South Africa, 4, 11, 12, 13, 19, 33
South America, vii, 1, 4, 8, 11, 15, 19, 43, 52, 66, 68, 80
South Pacific, vii, viii, ix, 37, 38, 41, 44, 45, 47, 52, 59, 61
Southeast Asia, vii, 1, 8
SP, 63, 138
space-time, 63
Spain, 12, 29, 52, 67, 88, 96, 98
species richness, 4, 19, 146
sperm, 11, 83
spin, 126
sponge, 72
Spring, 29, 78, 121
SS, 144
St. Petersburg, 31
stability, 39, 81, 85, 87, 139, 155
stabilization, 18, 52, 98
stabilizers, 95
starvation, 134
state(s), 9, 44, 70, 116, 139, 147
statistics, 119
steel, xi, 149, 150
sterile, 125
stoichiometry, 68
storage, 53, 70, 124
stratification, 44, 58, 62, 64
stress, 31, 80, 85, 87, 96, 99, 103, 138, 142, 145, 153
stressors, 20
stretching, 150
structure, 4, 18, 56, 81, 88, 92, 120, 138, 141
style, 99

substitution(s), 46, 67
substrate(s), 11, 12, 13, 14, 16, 26, 70, 85, 102, 138, 141, 148
succession, 138, 144, 146, 147, 148
suppression, 80
surfactants, 83
survival, 4, 11, 25, 46, 67, 80, 87, 89, 92
Sustainable Development, 34
Sweden, 52
Switzerland, 35
symptoms, 43, 49, 52, 54, 55, 56, 57
synchronize, 102
syndrome, 87, 99
synthesis, vii, x, 43, 77, 85, 123, 124, 125, 132, 133, 134, 154

T

Taiwan, 8
target, ix, 26, 77, 80, 154
taxonomy, 5
technical assistance, 119
techniques, 5, 26, 94
technological advances, 77
technologies, 66, 78, 90
teeth, 11
TEM, 126, 132
temperature, 10, 12, 13, 14, 15, 25, 26, 43, 44, 60, 66, 80, 84, 87, 88, 92, 99, 102, 105, 110, 114, 138, 142, 148, 154
temporal variation, 147
terrestrial ecosystems, x, 123
testing, 49, 65, 83
tetrachlorodibenzo-p-dioxin, 83
Thailand, 23, 34, 91, 98
thermal treatment, 25
thinning, 120
threonine, 52
tides, 6, 87
tin, 43, 104
tissue, ix, 5, 11, 14, 17, 19, 27, 28, 34, 43, 47, 48, 52, 53, 54, 59, 60, 65, 66, 74, 77, 79, 80, 85, 88, 95, 97, 99, 102, 106, 150, 151, 152, 153, 154
tourism, 40
toxic contamination, 17
toxic effect, 46, 52, 53, 54, 57, 89
toxic metals, 22, 34
toxicity, ix, 11, 21, 27, 28, 30, 38, 40, 43, 44, 47, 52, 53, 54, 55, 56, 57, 64, 65, 66, 69, 71, 74, 86, 90, 92, 135
toxicity equivalence factor, 52, 55, 56
toxicology, ix, 72, 73, 77, 90, 92, 95
toxin, ix, 6, 38, 40, 43, 45, 46, 49, 52, 53, 55, 56, 58, 61, 62, 63, 66, 67, 68, 69, 70, 72, 73, 74
trace elements, 20, 29, 94
trade, 25
traits, 40
transcription, 126
transcripts, 129
transformation(s), 44, 46, 47, 48, 49, 56, 58, 59, 64, 65, 67, 75
translocation, 17
transmission, 86, 127
Transmission Electron Microscopy, 126
transparency, 19
transplant, 118
transport, 20, 70, 121, 155
transportation, 25
treatment, 8, 19, 20, 25, 26, 29, 65, 82, 83, 115, 116
tremor, 54
triglycerides, 53
tumor, 52, 69
tumours, 80
turnover, 154
turtle, 18

U

U.S. Geological Survey, 97
UK, 2, 9, 56, 77, 94, 126
Ukraine, 8, 14
UN, 3, 23, 82
UNDP, 2
UNESCO, 32, 34, 67, 72
uniform, 141
Unionoida species, vii, 1
United, vii, 1, 2, 3, 6, 7, 8, 14, 16, 22, 23, 25, 28, 32, 33, 54, 92, 93, 99
United Kingdom, 2, 54, 93
United Nations, 2, 3, 23, 32
United Nations Development Program (UNDP), 2, 23, 32
United States, vii, 1, 3, 6, 7, 8, 14, 16, 22, 23, 25, 28, 33, 92, 99
urban, xi, 84, 134, 149
urine, 66, 79
Uruguay, 44, 138
USA, vii, 1, 3, 9, 22, 29, 31, 32, 34, 68, 72, 73, 74, 97, 125
UV, 95

V

validation, 28, 91

valve, 87, 94
variables, 40
variations, 44, 57, 79, 84, 87, 92, 116, 154
varieties, viii, 21, 23, 37, 38, 53
vector, viii, 38, 52, 54, 126
velocity, 7, 14, 118
Venus, 47, 48
vertebrates, 92, 124, 135, 156
vessels, 82
victims, 66
visualization, 125
vitamins, 6
vitellogenin, vii, 86, 124, 126, 132, 133, 134, 135
vomiting, 24, 43
vulnerability, 15

W

Washington, 121
waste, 3, 5, 7, 22, 84, 98
waste incineration, 22
wastewater, 13, 18, 20, 82, 83
water quality, viii, 2, 16, 17, 19, 20, 27, 65, 96, 141
weakness, 43
wealth, 6
weapons, 83
West Africa, 79
wetlands, 3
wetting, 83
WHO, 3, 22, 23, 24, 35
wild animals, 82
wildlife, 17, 27, 94, 135
wood, 14
World Bank, 23
World Health Organization, 3, 35
worldwide, viii, 2, 3, 4, 9, 20, 22, 25, 39, 40, 43, 51, 53, 63, 78, 79, 82, 86, 92, 150
wound infection, 24

Y

yellow fever, 82
yield, 14, 94
yolk, x, 86, 123, 124, 126, 127, 128, 132

Z

zinc, 81, 94, 156
zooplankton, 10, 15, 16, 81